PALEOCLIMATE

Princeton Primers in Climate

PALEOCLIMATE

Michael L. Bender

PRINCETON UNIVERSITY PRESS   *Princeton & Oxford*

Published by Princeton University Press, 41 William Street, Princeton,
New Jersey 08540
In the United Kingdom: Princeton University Press, 6 Oxford Street,
Woodstock, Oxfordshire OX20 1TW

press.princeton.edu

Library of Congress Cataloging-in-Publication Data
Bender, Michael L.
Paleoclimate / Michael L. Bender.
pages cm. — (Princeton primers in climate)
Includes bibliographical references and index.
ISBN 978-0-691-14554-9 (cloth) — ISBN 978-0-691-14555-6 (pbk.)
1. Paleoclimatology. I. Title.
QC884.B39 2013
551.609'01—dc23
2013011814

British Library Cataloging-in-Publication Data is available

This book has been composed in Minion Pro and Avenir LT Std

10 9 8 7 6 5 4 3 2 1

PALEOCLIMATE

Michael L. Bender

PRINCETON UNIVERSITY PRESS   *Princeton & Oxford*

Library of Congress Cataloging-in-Publication Data
Bender, Michael L.
Paleoclimate / Michael L. Bender.
pages cm. — (Princeton primers in climate)
Includes bibliographical references and index.
ISBN 978-0-691-14554-9 (cloth) — ISBN 978-0-691-14555-6 (pbk.)
1. Paleoclimatology. I. Title.
QC884.B39 2013
551.609'01—dc23
2013011814

British Library Cataloging-in-Publication Data is available

This book has been composed in Minion Pro and Avenir LT Std

10 9 8 7 6 5 4 3 2 1

Contents

Boxes

Preface

As part of the Princeton Primers in Climate Science series, *Paleoclimate* is a broad introduction to the subject for a scientifically literate audience, a reference for information about specific topics in the field, and a textbook for courses in climate and paleoclimate.

Earth's climate has undergone dramatic changes since early in the history of the planet. At one extreme, Earth was glaciated to the equator, more than once, for intervals that may have lasted millions of years. At another, climates were so warm that the Canadian Arctic was heavily forested and large dinosaurs lived on Antarctica. Four key factors have caused these climate modifications: changes in atmospheric greenhouse gas concentrations, changes in the amount of sun's radiation reflected directly back to space, changes in the position of the continents that guide winds and ocean currents, and changes in the brightness of the sun.

The first task of paleoclimate science is to identify, from observations of the geological record, the nature of past climate changes. The effort devoted to this task has been huge, and paleoclimate scientists have developed and used a very broad array of methods, some wonderfully imaginative. The second task is to use the resulting observations to synthesize a coherent, falsifiable narrative describing major paleoclimate events. The third task is to understand the dynamics shaping the events that are observed and described.

The motivation for these activities is twofold. First, climate history is a compelling topic that examines a fundamental feature of the environment. Second, observations of climate history help us to understand the range of possible climate responses to anthropogenic perturbations, and to test models simulating or predicting these responses.

This book describes the study of paleoclimate. The first chapter explains the main attributes of climate on the planet, including controls on global average temperature, patterns of winds and precipitation, and other first-order features of the environment. The book then describes seminal events in Earth's climate history. The starting point is the "faint young sun" problem: How could there have been water on Earth's surface early in the history of the planet, when the sun shone only about 70% as brightly as today? The next topic is "snowball Earth," periods before 600,000,000 years ago when the planet was glaciated to the equator, perhaps for millions of years. The subsequent chapter describes a paradigm that accounts for the regulation of greenhouse gases and Earth's temperature over the Phanerozoic Eon, the last, fossiliferous, 543 Myr (million years) of Earth history. Then comes the Late Paleozoic ice ages, an interval from about 360 to 270 Ma (millions of years before present) when Earth was periodically glaciated. That event was followed, from about 250 to 50 Ma, by very warm conditions, with forests covering the islands of the Canadian Arctic, for example. Around 50 Ma, Earth began cooling, a process that still continues. Superimposed on this long cooling were global climate cycles, whose magnitude and duration have changed with time. The last ice age ended 11,700 years ago, and we entered the Holocene, the interglacial period of the current glacial-interglacial climate cycle that hosted the development of civilization. Over the past two centuries, humans have become agents in the global climate system, most notably by emitting CO_2 (carbon dioxide) and other greenhouse gases.

These climate events played out in Earth's dynamic surface environment. Three attributes of this environment are particularly important. First, volcanism and other processes occurring in Earth's interior continuously release CO_2 to the atmosphere. This CO_2 warms the planet until it is removed by "weathering," the chemical reactions in which CO_2 is consumed while interacting with the crystalline rocks of Earth's surface. Second, the positions of the drifting continents establish a boundary condition for the climate system. Third, the evolving biota affect concentrations of greenhouse gases in the atmosphere, weathering reactions, and the reflectance of the planet.

Other than water, CO_2 is by far the most important greenhouse gas. Its atmospheric concentration is regulated by the balance between release from Earth's interior and removal by weathering. "Feedbacks" are interactions that depend on the state of a system. A positive feedback occurs when an increase in one property leads another property to change in a way causing the first property to increase further. A negative feedback occurs when a rise in the first property leads the second to change in a way that lowers the first property, this stabilizing the system. Feedbacks within the Earth system maintain the habitability of the planet while still allowing quite significant variations in the atmospheric CO_2 concentration and climate.

Continental positions continuously change as seafloor spreading moves the continents over the surface of the planet. Positioning of continents close to the equator during periods of "snowball Earth" is thought to have enabled the descent into static, cold climates. Continental positions have also contributed to the two great periods of oscillating ice ages, the Late Paleozoic ice ages (about 360–270 Ma) and the glacial climates of the past 34 Myr. Today, for example, glaciation is abetted by the presence of a continent centered over South Pole. Antarctica is permanently glaciated, leading to high reflectance of sunlight by the bright surface, and the presence of very cold waters in the Southern Ocean. There is also a large temperate and subpolar landmass in the Northern Hemisphere, on which ice sheets can grow and decay with a cycle time of 40–100 Kyr (thousand years).

Biota affect climate by producing and consuming CO_2 and other biogenic greenhouse gases that are important in Earth's heat balance. Plants produce O_2 and are the immediate source of that gas in air. They also help darken Earth's surface, thereby influencing the amount of the sun's heat that is absorbed by the planet rather than being reflected back to space. Finally, plants enrich soils in organic matter and hence in metabolic CO_2 as well, thereby facilitating the uptake of CO_2 by weathering. The biota thus influences background climate and is also implicated in several of the events of interest to us.

Life began by about 3.9 Ga (3.9 billion years ago); photosynthesis originated by 2.7 Ga, and the O_2 concentration of air rose to significant levels at about 2.4 Ga. Its rise would have led to the demise of greenhouse gases, possibly explaining Earth's first snowball event. Large changes in Earth's carbon cycle are associated with two snowball events occurring between 720–610 Ma. Plants colonized land around 400 Ma. They would have enhanced weathering, perhaps contributing to the Late Paleozoic ice ages that soon followed. Interactions between plants and climate have had an influence on more recent climate change that is important, if more subtle.

In this book, one chapter is devoted to each of the seminal climate events listed above. Each chapter describes the physical evidence for the nature of the event; presents a picture of the relevant climate cycles, where appropriate; and discusses hypotheses that have been advanced to explain the episode. The book aims to be accessible and concise rather than exhaustive, but summarizes viable competing hypotheses.

Paleoclimate is perhaps the oldest discipline in Earth science; it began in the nineteenth century, and earlier, with the debate about whether the surface environment of temperate areas was shaped by the biblical flood or by glaciers. By the middle of the twentieth century, many climate features associated with the recent ice ages had been identified. Progress accelerated dramatically after the Second World War with the advent of the methods of nuclear geochemistry, including radiometric dating and stable isotope geochemistry. Around this time, ships and tools for sampling seafloor sediments also advanced, leading to great improvement in our understanding of climate change in the oceans. Methods were also developed for recovering and studying cores drilled through the Greenland and Antarctic ice sheets. A seminal advancement in our understanding of the ice ages occurred when Swiss and French scientists learned to use these samples to characterize the CO_2 concentration of the past atmosphere. In the mid-1960s, the plate tectonics revolution led to a coherent understanding of the physical environment in which major climate changes occurred. Improvements in characterizing climate by geochemical and other tools, together with the use of simple and complex climate models to analyze observations, have advanced the field. At the same time, there has been something of a return to the roots of paleoclimate research, with a growth of interest in fieldwork leading to spectacular new insights. Finally, the challenges of understanding past climates, together with the growing awareness it offers of anthropogenic global change, have led to the recruitment of distinguished scientists from other fields who have made major contributions to the discipline. This constellation of resources and activity leads to the story in this book.

Acknowledgments

I AM GRATEFUL TO COLLEAGUES WHO READ DRAFTS OF part or all of this book, including Danny Sigman, Dan Schrag, Adam Maloof, David Lea, and especially Matt Huber, who caught major errors in the manuscript and was generous with his suggestions. Needless to say, any remaining errors are mine. Wally Broecker was my thesis advisor and continues to be an inspiration. I deeply appreciate the careful work of Sheila Dean, Natural Selection Editing and Research, who proofread and edited the manuscript. I also appreciate the encouragement and advice of my editor, Alison Kalett. Jenny Wolkowicki, production editor, helpfully made me toe the line, and Dimitri Karetnikov carefully worked with me on the figures. I always appreciate the love, companionship, and support of my wife and family.

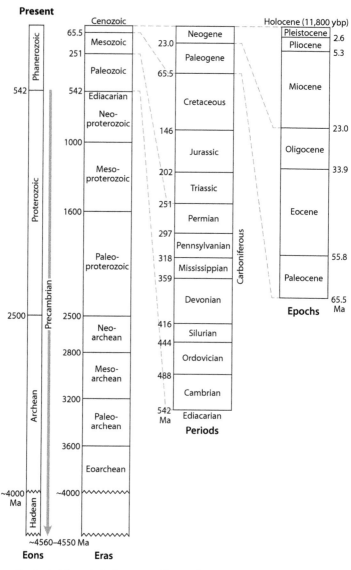

Geologic Time Scale (ages in millions of years before present)

1 EARTH'S CLIMATE SYSTEM

EARTH'S CLIMATE SYSTEM INCLUDES ALL THE REALMS of the planet that interact to produce the seasonal march of temperature, wind, and precipitation. Most important are the atmosphere; the oceans, including their linked chemical and biological processes; and the solid Earth insofar as it influences CO_2 concentration in air. Atmospheric processes govern climate over time scales of a few years or less. The oceans influence climate change over periods of decades to tens of millennia. Over periods of a hundred thousand years or more, interactions between the solid Earth and the surface environment fix the CO_2 concentration of air and Earth's average temperature.

In this chapter, we discuss the physical and chemical controls that determine the most important characteristics of Earth's climate. We start by discussing the decrease of pressure and temperature with elevation. We proceed to Earth's heat budget and the controls on global average temperature. We then examine the large scale circulation of Earth's atmosphere, and discuss how this circulation dictates prevailing wind directions at the surface and how it determines what regions of the globe get a lot of precipitation and what regions are dry. We discuss ocean circulation, biological processes in the ocean, and how these processes combine to change the partial pressure of CO_2 in the atmosphere over periods of decades to millennia. We end with a brief description of the geological processes that fix the average background CO_2 concentration of the atmosphere over hundreds of thousands of years.

ATMOSPHERIC PROPERTIES AND CLIMATE

Pressure and temperature as a function of altitude

Earth is heated by sunlight predominantly at ground level, which in turn warms the local lower atmosphere. Warm air expands and becomes buoyant, leading to vertical mixing. As air rises, it encounters lower pressures and expands into the void. Atmospheric pressure decreases with elevation in a way that reflects hydrostatic equilibrium in the atmosphere. In this condition, air is stabilized at a given altitude by the balance between gravity, which pulls the air mass down toward the surface, and the upward push exerted by the natural tendency of a gas to expand.

At a given elevation, pressure is simply the weight per unit area of the overlying column of air. This condition is expressed by the equation:

$$d\rho/dz = -g\rho = M_{air}/RT, \tag{1}$$

where ρ is density, z is height above the surface, g is gravitational acceleration, M_{air} is the molar mass of air

(29 gm/mole), R is the ideal gas constant, and T is Kelvin temperature. From this equation, one can show that pressure decreases by a factor of 1/e (0.37) for every ~7–8 km increase in elevation for typical air temperatures.

Temperatures are cold at higher altitudes because of the decrease of pressure with elevation. Consider a parcel of dry air large enough that it is not gaining or losing heat to the surrounding atmosphere. As this parcel rises, it encounters lower atmospheric pressure and "pushes out" into the surrounding air. In so doing, it uses energy, which leads it to cool. The cooling rate, or "lapse rate," is about 10°/km. If the air is wet, water condenses as it rises and reaches the dew point. Latent heat is released, and the lapse rate is smaller, typically 4–7°/km. Lower values correspond to warmer saturated air, with more water vapor and greater potential to release latent heat.

This decrease with temperature reverses at an altitude of about 11 km. The reversal is caused by the absorption of high energy (ultraviolet or UV) light from the sun due to reactions of the ozone cycle. These reactions are:

$$O_2 \rightarrow 2\,O \qquad\qquad (2)$$

$$O_2 + O \rightarrow O_3 \qquad\qquad (3)$$

$$O_3 \rightarrow O_2 + O \qquad\qquad (4)$$

$$O + O_3 \rightarrow 2\,O_2 \qquad\qquad (5)$$

Absorption of ultraviolet light by O_2 (oxygen) and O_3 (ozone) has the net effect of warming the surrounding air. It is this warming that causes the temperature to increase with altitude above about 11 km elevation. This increase in temperature continues to an elevation of ~50 km, where pressure is about 1% of the sea level value. Above 50 km, the temperature again begins to fall because O_2 is not abundant enough to allow significant O_3 production. There is an additional reversal at about 85 km elevation.

The *troposphere* is the atmospheric layer from the surface to the first temperature minimum, and the surface of minimum temperature is the *tropopause*. The *stratosphere* is the overlying layer of air from about 11 to 50 km elevation. These are the two lowest layers of the atmosphere.

Solar heating and radiative equilibrium

At the top of the atmosphere, the cross sectional area of Earth receives heat from the sun at the rate of 1368 watts m^{-2}. Spread over Earth's entire daytime and nighttime spherical surface, the average heating is 4 times lower, at 342 W m^{-2}. Some of this heat is reflected back to space. The remainder is redistributed in various ways between the land surface, ocean, and atmosphere. However, it is lost only by radiation of photons or electromagnetic radiation to space. The loss is described by the Stefan-Boltzmann equation:

$$\text{Rate of energy loss} = \sigma\, T^4, \tag{6}$$

where T is Kelvin temperature and σ is the Stefan-Boltzmann constant, 5.67×10^{-8} w m^{-2} K^{-4}.

Earth's average surface temperature changes only very slowly with time, so that heat must be lost at nearly the same rate at which it is received. Equation (6) can then be rearranged to solve for temperature. The temperature calculated for a heat flux of 342 W m^{-2} is 6°C, not too different from Earth's preindustrial average surface temperature of about 15°C. Unfortunately, there are two serious omissions in this calculation. The first becomes apparent by examining figure 1.1: much of the light reaching the Earth is simply reflected back to space, without ever warming the surface and contributing to Earth's heat budget. Sunlight is reflected by all surfaces, but the brightest (most reflective) are clouds, snow and ice, and deserts. Earth's global reflectance, or *albedo*, is 0.31. Correcting for albedo, a value of -19°C is calculated, which is $\sim$36°C too low for Earth's surface temperature. What's wrong?

Actually, nothing. A value of -19°C is Earth's radiative equilibrium temperature, but it is not achieved at the surface. The reason for this is the greenhouse effect, which is illustrated in figure 1.2. The top panel (a) shows the energy density of solar and Earth radiation as a function of wavelength. Wavelength increases to the right; frequency, and energy of electromagnetic radiation, increase to the left. Because the surface of the sun is so hot ($\sim$6000 K), most solar energy is radiated in the visible region of the electromagnetic spectrum. Radiation from the cool Earth, on the other hand, is in the longer wavelength, lower energy infrared region. Panels (b) and (c) show the fate of radiation as it passes through the

Fig. 1.1. Earth viewed from Apollo 17 in space. Landmasses from bottom: Antarctica (ice-covered), and Africa, with Madagascar to the east, and the Saudi Peninsula to the northeast. The Southern Ocean separates Antarctica from South Africa, the Atlantic is to the west of Africa, and the Indian to the east. Ice-covered Antarctica and clouds are the most reflective surfaces; the North African desert and Saudi Peninsula are next. The forests of tropical Africa are even less reflective, and the ocean is the least reflective realm. This image illustrates that much incoming sunlight (31% globally) is reflected back to space.

atmosphere. In these panels, white means that radiation is transmitted, and gray indicates that it is absorbed by interactions with molecules of the gases in air. Absorbed radiation is used to kick electrons into higher energy levels, and to increase vibrational and rotational frequencies of molecules. Panel (b) shows the absorption of radiation as a function of wavelength between the surface and the top of the atmosphere. Most solar radiation passes through the atmosphere "intact"; it is this property that allows us to view the sun, Moon, and stars. Most Earth radiation, on the other hand, is absorbed as it passes through the atmosphere. Absorption warms the air and leads to reradiation of the absorbed energy. Some of this reradiated energy is transmitted downward toward the surface, where it delivers an extra serving of heat. Thus, the surface is warmer than it would be if absorbing (or greenhouse) gases were absent from the atmosphere.

Panel (c) shows the fraction of radiation absorbed between 11 km and the top of the atmosphere. In this interval, almost all outgoing Earth radiation is transmitted, and there is no longer much warming of the atmosphere due to absorption of outgoing infrared radiation.

Ideally, there would be some level in the atmosphere below which infrared radiation is largely absorbed, and above which it is mostly transmitted. It is at this hypothetical level that Earth attains its radiative equilibrium temperature of $-19°C$. This level is at about 5 km elevation. The average lapse rate in the troposphere is about $6.5°C/km$, so that temperature rises by $33°$ from the

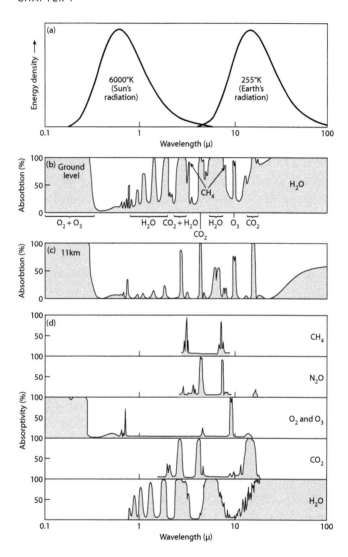

radiative equilibrium level to sea level. The calculated average sea level temperature is 14°C, close to the observed global average. The concept of a single level where the infrared radiation ceases is greatly simplified, but the idea is correct.

Panel (d) illustrates the absorption of radiation by different gases between the surface and the top of the atmosphere. In the ultraviolet, all absorption is due to O_2 and O_3, illuminating the role of ozone as the UV shield. In the infrared (IR), most absorption of radiation is due to water. Of the other so-called greenhouse gases, CO_2 is by far the most important absorber, followed by CH_4 (methane) and N_2O (nitrous oxide). Ozone also absorbs in the IR, and there is thus a small contribution to the greenhouse effect both from tropospheric and stratospheric ozone.

Fig. 1.2. (*a*) Energy density of radiation given off by black bodies at 6000 K (representing the sun) and 255 K (representing Earth) as a function of wavelength in micrometers. (*b*) Percentage of radiation absorbed while passing between the top of the atmosphere and the surface of the Earth. Most solar energy is transmitted, while most outgoing Earth radiation is absorbed. (*c*) Percentage of radiation absorbed between the top of the atmosphere and 11 km elevation. Most solar and Earth radiation is transmitted by the thin (and dry) atmosphere in this region. (*d*) Contributions of the different gases to absorption of radiation between the top of the atmosphere and the surface of the planet. Water is the most important greenhouse gas, followed by CO_2. From Peixoto and Oort 1992.

ATMOSPHERIC CIRCULATION

Sunlight warms Earth's surface, which in turn warms the atmosphere. Under these conditions, one might expect a meridional (latitudinal) circulation system, with warm air rising at the equator, and cool air sinking at the poles. There would be poleward flow aloft, and equatorward flow at the surface. In fact there are convection cells in the atmosphere, but they are not quite this grand. In the "Hadley cell," air rising at the equator flows to a latitude of about 30°, sinks to ground level, and flows back toward the equator. In the "Polar cell," air rises at about 60° latitude, flows toward higher latitudes, sinks at the poles, and again closes the loop by equatorward flow at the surface.

Because of an effect known as the Coriolis force, winds are westerly (from the west) in the upper, poleward flowing air of the Hadley cell. The air flowing poleward must, in the absence of friction, conserve its angular momentum (mass × velocity × radius). Air at the equator is moving toward the east at about the same rate that the surface is spinning on its axis. As this equatorial air rises and moves poleward, its eastward velocity increases in an absolute reference frame because the radius of the Earth is decreasing. With respect to the underlying ground, its velocity is increasing even faster, because the eastward velocity of the ground becomes smaller as the air moves to higher latitudes. From our perspective, with a reference frame rotating along with the Earth, an air mass traveling toward the pole will accelerate toward the east. The Coriolis force is the virtual force producing this acceleration.

Think about air at the equator in relation to the ground. In an absolute frame of reference, it is moving to the east at a velocity of 1671 km/hr, due to rotation of the solid Earth. In the absence of friction and turbulence, this air would be flowing to the east 484 km/hr faster than the ground when it reaches a latitude of 30°. In practice, there is such a feature in the upper troposphere: the jet stream, which flows eastward but at a lower velocity of 150–200 km/hr. Wind speeds must be low near the surface due to friction, and even above the surface, friction and other influences make wind speeds lower than those calculated when only considering conservation of angular momentum.

High velocities aloft, coupled with the decrease in temperatures between the Hadley cell and the Polar cell (at latitudes of roughly 30–60°), lead to a very different atmospheric circulation in these middle latitudes. Here, the circulation is much more chaotic, dominated by cyclones—large air masses rotating counterclockwise in the Northern Hemisphere and clockwise in the Southern Hemisphere. Cyclones travel to the east, leading to transitional regions known as fronts and the variable weather so characteristic of the midlatitudes. Along with the movement of cyclones, there are flows of warm air masses toward the poles and cool air masses toward the equator. These flows lead to the transport of heat from the tropics toward the poles, and attenuate the meridional temperature gradient. For dynamic reasons, there tends to be net sinking of air at the southern boundary of the midlatitude zone (around 30°) and rising air at the northern end (60°).

With this background it becomes fairly straightforward to understand the distribution of surface winds. In the region of the Hadley cell (0–30°), air rises at the equator. Aloft, it flows toward the pole and to the east. At ground level, there is a return flow. Winds turn toward the right (west) as they pass from a region where the surface is turning more slowly to a region where it is turning faster (the equator). The tropics are therefore regions of easterly winds (from the east) known as the trade winds. Between 30 and 60° latitude, there is a net westerly flow, upon which are superimposed the airflows associated with rotating cyclones. Hence, surface winds in this region are variable, despite the mean flow to the east. North of 60°, circulation of the polar cell is similar to that of the Hadley cell: air rises at the low-latitude boundary, flows toward the east aloft, sinks at the poles, and the surface return flow is again easterly.

This dynamic picture also explains the distribution of precipitation. Air rises at the equator and at 60° (on average), then sinks at around 30° and at the poles. When air rises, there are two changes that influence the degree of saturation of water vapor, and hence the amount of precipitation. First, rising air expands, decreasing the concentration of water vapor relative to the saturation concentration at which liquid will form. Second, rising air cools, lowering the equilibrium water vapor concentration. It turns out that the effects of cooling exceed those of expansion, and the degree of saturation of water vapor increases as air rises. Consequently, there is a belt of heavy precipitation along the equator. Sinking leads to the opposite effect: air warms, and its ability to hold moisture rises. This effect leads to the areas of low precipitation in the Northern Hemisphere around 30° N, accounting for the deserts of North Africa and the western United States, and dry climates elsewhere in the subtropics. In midlatitudes, fronts lead to rising air masses and abundant precipitation. Precipitation is low at very high latitudes because cold air can hold very little moisture.

Superimposed on these global patterns are important local features. Perhaps the most important are temperature and precipitation gradients over land produced by interactions between land and the nearby sea. Heat received by land is absorbed at the very surface and transmitted to depth by conduction, which operates very slowly and induces seasonal temperature cycles to depths of only a few meters. Heat received at the sea surface penetrates more deeply and is mixed rapidly to depths of tens to hundreds of meters. In other words, the land surface shares its heat to only a meter or so depth, while the ocean surface shares its heat to 100 m or more. Consequently, seasonal heating and cooling of land is much greater than that of the oceans. This feature manifests itself dramatically in two ways. First, in temperate and subpolar continental areas, seasonality is much stronger in the east than in the west. For example, Baltimore is only 2° further north than San Francisco, but its annual temperature cycle is four times larger (25°C range of monthly average temperatures compared with 6°). These zonal gradients are a sign of the prevailing westerly airflow in the middle latitudes. Air in San Francisco originates from the Pacific Ocean, and reflects its attenuated seasonal temperature cycle. On the other hand, air in Baltimore has crossed the continent and has acquired the large seasonal temperature fluctuations in the center of North America.

Another important local feature is the intense summertime heating of air over land. This makes air buoyant, causes it to rise, and draws in more wet air originating over the oceans. As air rises, it of course cools. Water condenses and falls as rain. This phenomenon gives rise to the monsoons, which involve intense summertime precipitation over large continental areas with meteorological links to the oceans.

THE OCEANS

Ocean circulation

The same laws of fluid flow govern the circulation of the oceans and the atmosphere. However, ocean circulation is very different from atmospheric circulation for two reasons. First, the oceans are heated from the top rather than from the bottom. Second, seawater is more dense than air; thus, ocean currents are much slower than winds. Like the atmosphere, the ocean is dynamic, although it mixes on far longer timescales (about one millennium vs. one year). Three factors cause the oceans to move or mix. First, waters that are more dense than their surroundings will sink, and more buoyant waters will rise. Second, winds transfer momentum to the sea surface, inducing lateral flow and in some cases vertical motions. Third, ocean tides and currents induce vertical mixing of waters in the ocean interior, especially over mountainous areas of the seafloor.

The deeper waters of the oceans originate as the densest waters at the ocean surface. The density of surface waters depends on both temperature and salinity. Surface waters of the oceans are heated in the low latitudes. In the high latitudes, they are cooled by interactions with the atmosphere during much of the year. Therefore, higher latitude waters are colder, as most of us have experienced, which increases their density. Surface waters are also subject to evaporation and precipitation. Evaporation removes water without removing salt, increasing salinity—and density. Precipitation has the opposite effect. Precipitation exceeds evaporation along the equator, and evaporation is more rapid in the subtropics (as on land, where the tropics are filled with rainforest, and the subtropics host the world's great deserts). Consequently, the salinity of seawater tends to be low on the equator, high in the subtropics (about 30° latitude), and lower toward the poles. Since salty water is denser than fresh, salinity variations raise the density of water in the subtropics and decrease it at low and high latitudes. Over the ambient range of surface ocean conditions, density changes due to temperature are roughly twice those due to salinity. Hence temperature wins the competition to influence density; the densest waters are found in the high latitudes, and it is here that waters sink to fill the deep oceans. Even in high latitudes, however, salinity has some impact. North Atlantic salinities are the highest of the polar regions, Southern Ocean salinities are lower, and North Pacific salinities are the lowest. It turns out that North Pacific densities are too low to allow deep water formation in this region. Consequently, it is the North Atlantic and Southern Ocean where waters form that fill the ocean basins below about 1000 m depth. For reference, the average deep ocean depth is ~3800 m.

Winds impart momentum to the sea surface, leading to the flow of water. The consequences are, however, somewhat surprising, because Earth's rotation leads flows to bend to the right in the Northern Hemisphere and to the left in the Southern Hemisphere. Winds blow from the east in the tropics and from the west at midlatitudes. These winds cause waters to flow to the north in the tropics and to the south in the midlatitudes. Waters thus "dome" in the center of the ocean basins, exerting a pressure gradient leading to a circular, or "gyre" flow. These waters circulate in a counterclockwise direction in the Northern Hemisphere and clockwise in the Southern Hemisphere.

Winds also join with density flow and interior mixing (or turbulence) to drive the exchange of waters between the surface and the abyss. Westerly winds blowing over the Southern Ocean drive the eastward-flowing Antarctic Circumpolar Current, which circles the continent. These also drive a flow to the left (north) in roughly the northern half of the Southern Ocean. This northward flow leads to a water deficit in the center of the basin, which induces the upwelling of waters from the deep ocean. Upwelling in turn activates the formation of new deep waters in the North Atlantic and the Southern Ocean. Also contributing to deep water formation is turbulent vertical mixing driven by tides and currents flowing over rough bottom topography. In this process, heat is mixed from shallow depths into the deep ocean. The warmed, buoyant, and deep waters flow toward the surface, causing more deep waters to form.

Superimposed on the large-scale, annual mean flows are seasonal cycles associated with warming or cooling of surface waters. These seasonal cycles are intense in the upper 50 m or so, and have a significant imprint to depths of 100 m, and much more in polar regions. In summer, there is a thin mixed layer at the surface, typically 20–50 m deep, with waters below cooling progressively to annual average values. In wintertime, surface waters are cooler and denser. They thus mix readily with deeper waters. Vertical mixing, and the transfer of waters from the surface to the ocean interior, are thus predominately a wintertime process.

The conflation of mixing processes leaves the oceans with three great realms. There is a warm surface layer extending from about 45° N to 45° S, and from the surface to about 100 m depth. There is a cold water realm (temperature <4°C), extending from the Arctic to Antarctica, and up from the seafloor. The cold water realm includes the surface ocean at latitudes poleward of about 60°. From there to the midlatitudes, its upper boundary progressively deepens to about 1000 m depth, which is typical for the region between 45° N and S. Finally, there is an intermediate realm, which shows a high seasonality at the surface, from ~45° to 60°. As in the polar realm, subsurface waters form at the surface during wintertime, and progressively cooler waters form at progressively higher latitudes. Wintertime surface waters sink to moderate depths to fill the "main thermocline" lying within this intermediate realm and extending from ~200 to 1000 m depth, and from about 60° N to 60° S.

Ocean Biogeochemistry

The interaction between ocean chemistry and biology reflects five generalizations or facts about the ocean. First, all photosynthesis takes place in the sunlit upper layer of the oceans, which is ~100 m deep. Second, almost all metabolism in the oceans is by prokaryotes and single celled eukaryotes. Third, most organic matter is heavier than water and tends to sink. Fourth, almost all organic matter is eventually respired or remineralized (metabolized back to inorganic constituents) rather than preserved in deep-sea sediments. Finally, ocean currents transport dissolved chemicals in the direction of flow, while turbulence mixes waters with higher and lower concentrations and attenuates concentration gradients.

These generalizations explain the basic cycles of biologically active chemicals and their distributions in the oceans. In the upper ocean, single-celled plants (phytoplankton) assimilate carbon, nitrogen, phosphorus, trace metals, and other elements to make tissue. Most of this tissue is rapidly remineralized (transformed to inorganic chemicals by respiration), with the consumption of O_2 and the release of dissolved inorganic carbon, nutrients, and metals. A fraction survives to sink toward the sea-floor, and this component is mostly remineralized as it sinks. The process depletes shallow waters in biologically active elements, and enriches subsurface waters. If the oceans were stagnant, nutrients would be completely drained from the sunlit zone and life in the surface would cease. However, upwelling and turbulent mixing return nutrients to the surface, where they are again assimilated by organisms.

The most interesting elements in this process are those in shortest supply relative to the biological demand. These elements, especially nitrogen (N), phosphorous (P), silica (Si), and iron (Fe), are almost completely removed from the sunlit zone over much of the oceans. The exception is in areas where subsurface waters are rapidly upwelling back to the surface. The most important site for this process is in the Southern Ocean, and also in the tropics, though to a much lesser extent. There is a plentiful and steady supply of nutrients in subpolar regions where deep waters mix to the surface during wintertime. In the subtropics, slow diffusion through the thermocline maintains a moderate supply of nutrients to the sunlit waters.

These three sequential processes—assimilation of biologically active elements, sinking of surface waters, and remineralization in the dark ocean—determine the distribution of biologically active elements in the sea. These elements are depleted in surface waters but enriched in deep waters. Remineralization continues along the route of deep water flow, and waters become progressively more enriched in nutrients as they flow from the deep Atlantic to the deep Pacific.

Dissolved inorganic carbon (DIC, the sum of CO_2, HCO_3^- [bicarbonate ion], and CO_3^{2-} [carbonate ion]) in the oceans is obviously utilized by biological activity, but it is never depleted by more that about 10%. There simply is not enough N and P in seawater to support more carbon uptake. Nevertheless, this degree of nutrient utilization has important consequences for atmospheric CO_2. Of the ~40,000 Gt (gigatons) of carbon in "mobile reservoirs" on Earth's surface, about 1.5% is in the atmosphere as CO_2, about 3% is in the land biosphere and soils, and the lion's share is dissolved in the oceans as DIC. The concentration of DIC in seawater, and the pH (or a related property), determine the partial pressure of CO_2 in surface seawater. Since the amount of DIC in the oceans is so much greater than the atmospheric CO_2 inventory, the partial pressure of CO_2 in surface seawater sets the concentration of CO_2 in the atmosphere.

The interplay of three processes can cause changes in the partial pressure of CO_2 (pCO_2) at the sea surface, and in air, over timescales of 10^2–10^4 years. The first process is biological utilization of DIC, which lowers pCO_2. The second is production of skeletal calcium carbonate ($CaCO_3$), which raises pCO_2 by converting 2 HCO_3^- ions into 1 CO_3^{2-} ion and one CO_2 molecule. The third is the riverine input of HCO_3^- and the burial of skeletal $CaCO_3$ in deep-sea sediments. These processes play the fundamental role in glacial-interglacial CO_2 variations.

Over longer timescales other processes dominate, as discussed next.

REGULATION OF ATMOSPHERIC CO_2 AND EARTH'S TEMPERATURE OVER MILLIONS OF YEARS

There is a fairly simple hypothesis for the regulation of Earth's average temperature. It invokes our understanding that volcanism, and other degassing processes associated with Earth's hot interior, steadily add CO_2 to the atmosphere. At the same time, weathering removes CO_2 to balance this input. *Weathering* is the attack of carbonic acid on rock-forming minerals of the solid Earth. In this process, rock-forming minerals are degraded to clay minerals, which are depleted in SiO_2 (silica dioxide) and cations, and CO_2 is converted to HCO_3^-. The dissolved products go into groundwater and eventually to the ocean. An example is the weathering of potassium feldspar to kaolinite:

$$4\,KAlSi_3O_8 + 4\,CO_2 + 6\,H_2O \rightarrow Al_4Si_4O_{10}(OH)_8 + 4\,K^+ + 4\,HCO_3^- + 8\,SiO_2 \tag{6}$$

If Earth's climate is stable, CO_2 input to the atmosphere by volcanism must be nearly in balance with CO_2 removal by weathering. A simple feedback maintains this balance. If CO_2 input is faster than consumption by weathering, the CO_2 concentration of air rises, and temperature warms. Under these conditions, weathering will accelerate, mainly because chemical reactions speed up as temperature rises. If volcanic input slows, CO_2 will fall, temperatures will cool, weathering will slow, and CO_2 input and output will once again come into balance.

Weathering is very slow below the freezing point, and chemical reactions typically double in rate for a 10° C rise in temperature. Therefore it is possible to compensate for relatively large changes in CO_2 outgassing with relatively modest changes of temperatures.

This idea was originally developed by Walker et al. (1981), formalized into a mathematical model by Berner et al. (1983), and modified and refined by Berner and colleagues in subsequent papers (Berner 2006; Berner and Kothavala 2001). Berner's recent models attempting to explain atmospheric CO_2 changes invoke many other important processes influencing the atmospheric CO_2 balance. Mathematical descriptions of the carbon cycle focus on the past 543 Myr (the Phanerozoic), and we will discuss this work further below.

IMPLICATIONS FOR PALEOCLIMATE

The discussion in this chapter illustrates that there are three reasons that Earth's average temperature might vary. First, the sun could have been shining more or less brightly in the past. Second, the concentration of greenhouse gases could have been higher or lower. Third, Earth's albedo may have changed. The study of paleoclimate involves characterizing Earth's climate in former times, and identifying properties that allow us to distinguish between these three possibilities in order to explain the dynamics of major climate changes of the past.

REFERENCES

Berner, R. (2006), Geocarbsulf: A combined model for Phanerozoic atmospheric O_2 and CO_2, *Geochimica et Cosmochimica Acta*, *70*, 5653–5664.

Berner, R., and Z. Kothavala (2001), Geocarb III: A revised model of atmopsheric CO_2 over Phanerozoic time, *American Journal of Science*, *301*, 182–204.

Berner, R., A. Lasaga, and R. Garrels (1983), The carbonate-silicate geochemical cycle and its effects on atmospheric carbon-dioxide over the past 100 million years, *American Journal of Science*, *260*, 641–683.

Peixoto, J. P., and A. L. Oort (1992), *Physics of Climate*, New York: American Institute of Physics.

Walker, J.C.G., P. B. Hays, and J. F. Kasting (1981), A negative feedback mechanism for the long-term stabilization of Earth's surface temperature, *Journal of Geophysical Research–Atmospheres*, *86*, 9776–9783.

2 THE FAINT YOUNG SUN

EARLY IN EARTH'S HISTORY, THE SUN'S ENERGY OUT-
put was about 30% less than at present. Temperature
scales with energy flux to the one-quarter power. Early
Earth temperature would thus have been lower than at
present by about 8.5%, falling to 264 K ($-9°$C) if the
greenhouse effect raises Kelvin temperature by a constant fraction. Under this condition, Earth would have
been frozen over. However, there is plenty of evidence
for liquid water on the surface of ancient Earth. The
solution to this conundrum appears straightforward:
atmospheric CO_2 simply rose to some high level at
which Earth's temperature was similar to today's. At
that point, weathering would have balanced the volcanic CO_2 input, as discussed at the end of chapter 1.
However, this explanation is problematic, because the
chemical composition of ancient sediments and soils
was apparently incompatible with such high CO_2 levels.
The alternatives are that ancient Earth had an enhanced
greenhouse effect from either clouds or elevated concentrations of other greenhouse gases, or that the planet's albedo was much lower early in its history. Feulner
(2012) recently reviewed this problem.

THE FAINT YOUNG SUN PROBLEM

The solar system accreted by the gravitational collapse of a mass of dust and gas in our region of space. The oldest minerals on Earth, found in meteorites, go back to 4.567 Ga (billions of years before present). Accretion of Earth took 50–100 Myr (million years). By about 4.52 Ga, Earth had grown to approximately 80% of its present mass, its core was largely intact, it had partly differentiated into a chemically distinct mantle and crust (Boyet and Carlson 2005), and a giant impact had recently lead to the formation of the Moon (Canup 2004). There is evidence from lunar and Martian meteorites and micrometeorites that Earth was heavily bombarded by large bodies until about 3.9 Ga.

The earliest evidence of water comes from studies of zircons in Australian granites dating to about 3.8 Ga. These rocks contain relict grains of the minerals zircon ($ZrSiO_4$) dating back as far as 4.3 Ga. In some of the zircon crystals, the ratio of the heavy to the light stable isotope of oxygen, $^{18}O/^{16}O$, indicates that these minerals incorporated water present at the Earth's surface (Mojzsis et al. 2001; Trail et al. 2007). By 3.8 Ga, photosynthesis left its signature in the isotopic composition of sedimentary organic carbon (Nisbet and Nisbet 2008), again pointing to the presence of liquid water on the planet. Finally, by 3.5 Ga, rocks originating as sediments had formed, and are found in many places. The oldest known glacial deposits date to about 2.5 Ga (chapter 3). It thus appears that a climate similar to today's, or warmer, was common even when Earth was young.

This situation is surprising because the sun shone less brightly earlier in its history. At the time it first accreted, the sun was composed of about 75% H (hydrogen) and 25% He (helium), with trace amounts of other elements. It contracted and heated to the point where, in its core, 4 atoms of H fused together in a series of reactions to form one atom of He. The mass of He is smaller than the mass of 4 H atoms. The lost mass is converted to thermal energy, and is the ultimate source of heat or light given off by the sun. As the sun ages and the "burning" of hydrogen progresses, the H content of the sun's core diminishes. The core responds by contracting, leading to heating. Thus, paradoxically, fusion accelerates as hydrogen fuel is depleted. For this reason, solar luminosity increases with time, at least until all the H in the core is converted to He (at which point the sun will be about 10 Ga in age).

According to the standard model of the sun, solar luminosity very early in Earth history was 0.71 times as bright as today, 80% as bright at 2.8 Ga, and 95% as bright at the dawn of the Phanerozoic (543 Ma) (Gough 1981).

KEEPING EARTH WARM DESPITE A DARKER SUN

The question thus arises as to how there could have been liquid water on the planet early in Earth history. A simple answer is provided by the schematic view of global temperature and CO_2 regulation presented in chapter 1; CO_2 will accumulate in air from volcanic emissions and other processes involving outgassing until temperature warms enough for weathering to balance the input. This idea seems to fail, however, because of two indications that the atmospheric CO_2 concentration was too low to sufficiently warm the planet (Rye et al. 1995). First, the absence of siderite ($FeCO_3$) indicates that the CO_2 level was not high enough for this mineral to form (Rosing et al. 2010; Rye et al. 1995). Second, an analysis of the mass balance of soils leads to the same conclusion (Sheldon 2006); weathering was much less intense than would be expected if CO_2 alone were responsible for warming. Apparently some other process must have helped warm the planet, although we need to be mindful that there are large uncertainties associated with both of these approaches to constraining ancient concentrations of atmospheric CO_2.

Methane is a greenhouse gas that today is produced biologically and destroyed mainly when oxidized by O_2. For three reasons, it is an attractive candidate for an enhanced early Earth greenhouse effect. First, it is thought that the O_2 concentration of air was very low for most of the Precambrian. In the absence of O_2, organisms turn to methanogenesis for metabolic energy. In methanogenesis, organisms produce methane by decomposing organic matter to CH_4 and CO_2. Second, earlier in its history, the planet would have been more tectonically active, and CH_4 would have been released more rapidly from Earth's interior. Third, today atmospheric CH_4 is destroyed by O_2 after about a decade. In the absence of O_2, the residence time would be much longer, and atmospheric CH_4 would build up to a much higher concentration. Then there is a bonus: when methane concentrations are sufficiently high, the photolysis of this gas in air leads to the formation of significant amounts of C_2H_6 (ethane), which is itself a powerful greenhouse gas (Haqq-Misra et al. 2008).

Figure 2.1 explains how global temperature can change as a function of CO_2 concentration with selected levels of atmospheric CH_4 (augmented by ethane, C_2H_6). According to this diagram, for example, if the CH_4 concentration is 10^{-3} atmospheres ($\sim 10^3$ times higher than the preindustrial value), global temperature would be about 276 K (3°C) when $CO_2 = 10^{-3}$ atmospheres (nearly 4 times the preindustrial value). This figure illustrates three constraints on possible solutions for an ice-free planet. First, the absence of siderite ($FeCO_3$) puts an upper limit on the atmospheric CO_2 concentration. Because of chemical equilibrium, this limit increases with Earth's surface temperature (Rye et al. 1995). So the CO_2 concentration must lie to the left of the "siderite" line in figure 2.1. Sheldon (2006) has argued, based on the intensity of weathering in paleosols (ancient soils), that the correct upper limit for pCO_2 would be somewhat lower, that is, 23 times present. Second, since the planet was unglaciated, Earth's average temperature must have been greater than 273 K, and probably higher than today's average global temperature of 289 K. Third, the CO_2

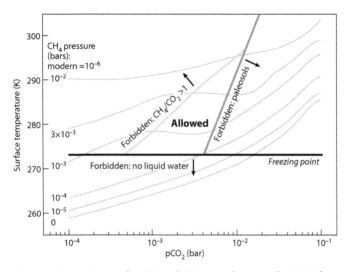

Fig. 2.1. Limitations to the CO_2 and CH_4 greenhouse early in Earth history (Haqq-Misra et al. 2008). The y-axis is average global temperature, and the x-axis is pCO_2 in bar or atmosphere (preindustrial $= 280 \times 10^{-6}$). The plot includes the greenhouse effect from C_2H_6 (ethane), which forms from photochemical reactions involving CH_4 (methane). The solid lines show the variation between temperature and pCO_2 for different levels of the atmospheric CH_4 concentration. The region where $CH_4/CO_2 > 1$ is forbidden because a reflective white haze would form, cooling the planet. The region to the right of the "paleosols" line is forbidden because the sediment and soil properties indicate CO_2 never rose above the relevant levels. The region below the "no liquid water" line is forbidden because, contrary to observations, the planet would be frozen. The lower part of the "Allowed" region is improbable because temperatures would still be low enough to cause extensive glaciation. There is only a very small area where temperatures are higher than today, a result in conflict with an ice-free Earth. Therefore, apparently other contributors to planetary warmth existed, including other greenhouse gases (possibly OCS), a cirrus cloud greenhouse, or a lower albedo early in Earth history.

concentration must be greater than the CH_4 concentration. When CH_4 is greater, its polymerization leads to the formation of a white haze that reflects sunlight and actually cools the planet, an effect opposite to the one required. Given these constraints, CO_2 and CH_4 concentrations could both have been high enough to contribute to significant greenhouse warming. However, if we assume that the average temperature of the ice-free Earth was at least 293 K (20°C), no allowable combination of CO_2 and CH_4 concentrations leads by itself to an unglaciated planet, and we need to look for other sources of warming as well.

There are three candidates. First, a provocative recent paper argues for a major role of OCS (carbonyl sulfide) in the Precambrian greenhouse. Today, this gas is supplied to air primarily by microbial activity in the oceans, and is removed by oxidation with a residence time of about five years. Its budget in the Precambrian is unknown, but the gas would be more stable in the O_2-free atmosphere. Ueno et al. (2009) concluded that OCS had an abundance of 5–10 parts per million (ppm) in the atmosphere prior to about 2.5 Ga. Their evidence comes from a distinctive pattern in the relative abundance of three stable sulfur isotopes (^{32}S, ^{33}S, and ^{34}S; box 1) in rocks older than 2.5 Ga. The distinctive pattern of sulfur isotope abundance before 2.5 Ga signifies the penetration of UV light to Earth's surface, and hence the absence of O_3, prior to 2.5 Ga. The details of the sulfur isotope abundance patterns serve as a fingerprint pointing to OCS concentrations of 5–10 ppm, a level associated with a significant

Box 1
Stable Isotope Abundance Variations

Most elements have two or more stable isotopes, which differ from each other in the number of neutrons in the nucleus. Stable isotopes are denoted by the chemical signal (e.g., O for oxygen) preceded by a superscript equal to the sum of the number of neutrons and protons. Thus, for example, ^{16}O indicates an oxygen atom whose nucleus has 8 protons and 8 neutrons, while ^{18}O indicates an oxygen atom with 8 protons and 10 neutrons. Stable isotopes have the same number of protons and electrons, and hence similar chemistry. In physical processes, such as evaporation, and also in chemical processes, such as the hydration of CO_3^{2-} to make HCO_3^-, isotopes behave almost identically; after all, they have the same number of protons in the nucleus, and are the same element. However, there are slight differences in their behavior due to the differences in mass. The consequences are most easily understood for the physical processes. Thus, for example, $H_2^{16}O$ is lighter than $H_2^{18}O$, and evaporates faster.

The reason for fractionation of the isotopes during chemical transformations is a little more complex. Molecules with heavy atoms vibrate more slowly, are more stable, and form bonds that are harder to break than molecules with light atoms. In kinetic processes (such as photosynthesis and respiration), molecules with light atoms react faster, and reaction products tend to be enriched in light atoms. In equilibrium processes, there can be significant isotope fractionations because the inclusion of heavy atoms favors some compounds more than others. An important example is the equilibrium of CO_3^{2-} with seawater; the heavy oxygen atom, ^{18}O, is enriched in the CO_3^{2-} by about 3% relative to its abundance in water. The magnitude of the enrichment is temperature dependent, giving a clue as

(Box 1 continued)

to how the oxygen isotope composition of $CaCO_3$ is relevant to paleoclimate (Urey 1947). Variations in the stable isotope abundance of hydrogen, carbon, nitrogen, oxygen, and sulfur give important information about geochemistry, biogeochemistry, and paleoclimate. The application of stable isotope geochemistry is ubiquitous over a very broad range of earth science studies.

Because isotopes are, after all, forms of the same element, the variation in their abundance is small. Variations in isotope ratios of 1% or so are common; 0.001% may be measurable and significant. Isotope abundance variations are precisely measured by comparing the ratio in a sample to that in a reference, using a mass spectrometer specifically designed for this purpose.

Variations in isotope abundances are reported in the δ notation, as the difference, in parts per thousand, or per mil, between the isotope ratio in a sample and in a reference. For example, there are two carbon isotopes, ^{12}C (6 protons and 6 neutrons) and ^{13}C (6 protons and 7 neutrons). The isotope ^{12}C comprises nominally 98.9% of C atoms and ^{13}C comprises 1.1%. The $^{13}C/^{12}C$ ratio is expressed as $\delta^{13}C$, where $\delta^{13}C$ has units of ‰, called "per mil."

$$\delta^{13}C = [(^{13}C/^{12}C)_{sample} / (^{13}C/^{12}C)_{reference} - 1] \times 10^3 \ (‰)$$

Hydrogen has two stable isotopes, ^1H and ^2H. For historical reasons, ^2H is called deuterium (D). The D/H ratio is about 0.015%. Variations in the D/H abundance are expressed as δD:

$$\delta D = [(D/^1H)_{sample} / (D/^1H)_{reference} - 1] \times 10^3 \ (‰)$$

Nitrogen has two isotopes, ^{14}N (99.626%) and ^{15}N (0.374%). $\delta^{15}N$ is:

$$\delta^{15}N = [(^{15}N/^{14}N)_{sample} / (^{15}N/^{14}N)_{reference} - 1] \times 10^3 \ (‰)$$

Oxygen has three isotopes: ^{16}O (99.76%), ^{17}O (0.04%), and ^{18}O (0.2%). $\delta^{18}O$ is:

$$\delta^{18}O = [(^{18}O/^{16}O)_{ref} / (^{18}O/^{16}O)_{sample} - 1] \times 10^3 \, (‰)$$

and $\delta^{17}O$ is expressed analogously.

Finally, sulfur has four stable isotopes: ^{32}S (95.0%), ^{33}S (0.8%), ^{34}S (4.2%), and ^{36}S (0.02%). The ratio $\delta^{34}S$ is:

$$\delta^{34}S = [(^{34}S/^{32}S)_{sample} / (^{34}S/^{32}S)_{ref} - 1] \times 10^3 \, (‰).$$

Both $\delta^{33}S$ and $\delta^{36}S$ are defined analogously.

Isotope abundance variations are generally expressed as the ratio of a minor isotope to the most abundant isotope. Changes in the relative abundance of stable isotopes are independent of the absolute concentrations. Whether a sample has an $^{18}O/^{16}O$ ratio of 1 or 1/500 (the approximate ratio in nature), $H_2^{18}O$ will always evaporate about 7‰ faster than $H_2^{16}O$, and at room temperature the $^{18}O/^{16}O$ ratio of CO_3^{2-} will still be 30‰ higher than the ratio in water.

REFERENCE

Urey, H. C. (1947), The thermodynamic properties of isotopic substances, *Journal of the Chemical Society*, 562–581. doi: 10.1039/JR9470000562.

greenhouse. The disappearance of the distinctive sulfur isotope pattern at about 2.5 Ga then signals the appearance of O_2 in air at significant levels.

The second possibility for an enhanced greenhouse comes from cirrus clouds in the tropics. Cirrus clouds are thin and "wispy." Like all clouds they both cool and warm the Earth; their reflectivity is important in the planetary albedo, while their absorption of outgoing long-wave radiation leads to a significant greenhouse effect. For most types of clouds, cooling due to albedo exceeds warming due to the greenhouse effect. However, cirrus clouds are unusual in that their greenhouse effect may exceed albedo, leading to a net warming. They are currently common in the tropics, in part because they form from moisture raised to high altitudes by cumulus towers. Rondanelli and Lindzen (2010) have shown that increased cirrus clouds in the tropics could warm the planet by significantly strengthening its greenhouse. The amount of warming achieved by increasing tropical cirrus depends on various assumptions; a reasonable estimate is about 5°C. This level of warming would not be enough to compensate for the cooler sun, but would contribute in addition to other influences, or "forcings."

Rosing et al. (2010) proposed a third solution to the faint young sun problem by suggesting that the planetary albedo was much lower early in Earth history. They put forward a number of factors that would lead to this difference. First, the continental surface area would have been free of vegetation, with a higher albedo than today. However, continents may have been smaller early in Earth history, leading to a net decrease in the continental contribution to the planetary albedo. Second, cloud condensation nuclei would have been rarer, cloud water droplets larger, and the cloud albedo correspondingly lower. Cloud droplets generally condense around salt nuclei originating largely from biogenic emissions. Most important are those of eukaryotic algae, which only evolved around 2

Ga. Fewer cloud condensation nuclei would translate to larger droplets and lower reflectance, just as coarse sugar is darker than fine-grained confectioner's sugar. Rosing et al. (2010) calculated that the lower albedo could have warmed the planet by up to 10°C. They postulated that an early Earth kept above freezing by a CO_2 level modestly higher than today (900 ppm vs. a preanthropogenic level of 280), a far higher CH_4 concentration (900 compared with 0.7), and a planetary albedo as low as 0.2, would lead to global average temperatures of about 5°C.

In summary, high concentrations of CO_2, CH_4, OCS, and C_2H_6 (derived from atmospheric reactions of CH_4) could all contribute toward a greenhouse effect sufficient to overcome the faint young sun and maintain moderate climates during the Precambrian. The composition of soils suggests that CO_2 was not elevated sufficiently to explain, by itself, global temperatures above freezing. There are reasons to believe that the CH_4 concentrations were dramatically elevated, though no physical evidence has yet been found for this. There are also reasons to believe that OCS concentrations were far higher than today; remarkably, the isotopic composition of sedimentary sulfate minerals argues for OCS concentrations elevated to the point where this gas made a significant contribution to greenhouse warming. Increased cirrus cloud in the tropics, and a lower planetary albedo resulting from a smaller continental area and fewer cloud condensation nuclei, could have combined with elevated greenhouse gas levels to maintain a large liquid ocean, rather than a frozen planet.

REFERENCES

Papers with asterisks are suggested for further reading.

Boyet, M., and R. Carlson (2005), ^{142}Nd evidence for early (>4.53 Ga) global differentiation of the silicate Earth, *Science*, *309*, 576–581.

Canup, R. (2004), Dynamics of lunar formation, *Annual Reviews of Astronomy and Astrophysics*, *42*, 441–475. doi: 10.1146/annurev.astro.41.082201.113457.

*Farquhar, J., H. Bao, and M. Thiemens (2000), Atmospheric influence of Earth's earliest sulfur cycle, *Science*, *289*, 756–758.

Feulner, G. (2012), The faint young sun problem, *Reviews of Geophysics and Space Physics*, *50*. doi: 10.0129/2011RG000375.

Gough, D. O. (1981), Solar interior structure and luminosity variations, *Solar Physics*, *74*, 21–34.

*Haqq-Misra, J., S. Domagal-Goldman, P. Kasting, and J. Kasting (2008), A revised, hazy methane greenhouse for the Archean Earth, *Astrobiology*, *8*, 1127–1137.

Mojzsis, S., T. Harrison, and R. Pidgeon (2001), Oxygen-isotope evidence from ancient zircons for liquid water at the Earth's surface 4,300 Myr ago, *Nature*, *409*, 178–181.

Nisbet, E., and R. Nisbet (2008), Methane, oxygen, photosynthesis, rubisco and the regulation of the air through time, *Philosophical Transactions of the Royal Society B-Biological Sciences*, *363*, 2745–2754.

Rondanelli, R., and R. Lindzen (2010), Can thin cirrus clouds in the tropics provide a solution to the faint young Sun paradox? *Journal of Geophysical Research–Atmospheres*, *115*. doi: 10.1029/2009JD012050.

Rosing, M., D. Bird, N. Sleep, and C. Bjerrum (2010), No climate paradox under the faint early Sun, *Nature*, *464*, 744–747.

Rye, R., P. Kuo, and H. Holland (1995), Atmospheric carbon dioxide concentrations before 2.2 billion years ago, *Nature*, *378*, 603–605.

Sheldon, N. (2006), Precambrian paleosols and atmospheric CO_2 levels, *Precambrian Research*, *147*, 148–155.

Trail, D., S. Mojzsis, T. Harrison, A. Schmitt, E. Watson, and E. Young (2007), Constraints on Hadean zircon proto-liths from oxygen isotopes, Ti-thermometry, and rare earth elements, *Geochemistry, Geophysics, Geosystems*, *8*. doi: 10.1029/2006GC001449.

Ueno, Y., M. Johnson, S. Danielache, C. Eskebjerg, A. Pandey, and N. Yoshida (2009), Geological sulfur isotopes indicate elevated OCS in the Archean atmosphehre, solving faint young sun paradox, *Proceedings of the National Academy of Sciences of the United States*, *106*, 14784–14789.

As indicated in the previous chapter, it seems that ice sheets were absent during most of Precambrian time. The Precambrian is characterized by abundant sediments that formed in a marine environment. Limestones, for example, that today are most common in low latitudes, are abundant in Precambrian sequences. Nevertheless, there is evidence for glaciation at several times during the Precambrian, including four spectacular events in which the extent of glaciers was far greater than in more recent times. The earliest known glaciation is dated to 2.9 Ga, recorded in sediments from the Mozaan Group of the Pongola Supergroup in eastern South Africa. Evidence includes the presence of diamictites. These are deposits of rock and dirt that appear to be of glacial origin; they include a wide range of rock types, angular rocks (not rounded in streambeds), and a large range of particle sizes. Some rocks have grooves that were apparently carved by stones dragged along the base of the glacier. There are also dropstones, rocks within layered sediments that were released by melting icebergs.

Evidence exists for two modes of Precambrian glaciation after 2.9 Ga. The first is cycles of sediment types, or "facies," that characteristically form under waters of specific depths. The sequence of facies thus may reflect a changing sea level, which would itself record growth of ice sheets (falling sea level) or their decay (rising sea level). Grotzinger (1986), for example, found sections in the Rocknest Formation of northwestern Canada with many such changes, and concluded that they reflected repeating glacial cycles. Interpreting cyclic sediments as sea level changes is, however, controversial because the facies changes may result from the random migration of beaches, stream channels, and coastal ponds.

The second mode of Precambrian glaciation is quite extraordinary: there is evidence that, in four different "snowball" episodes, Earth was glaciated to sea level in the tropics. These events are dated to 2.4 Ga, 1.9 Ga, 0.7 Ga, and 0.63 Ga. At each time, sediments are present with properties that reflect a glacial origin. There are also sedimentary or volcanic rocks present with magnetic properties suggesting that they formed at low latitudes. The nature of snowball events is highly controversial, but there is compelling physical evidence for extensive glaciation unlike anything seen since the beginning of the Phanerozoic, 543 Myr ago. The snowball glaciations are thus the focus of this chapter.

EVIDENCE FOR SNOWBALL GLACIATIONS

Some snowball deposits indicating a glacial origin are similar to those recognized for much more recent times. They include rocks whose surfaces have been polished as ice moved over them, rocks with grooves carved by stones dragged along the base of the glacier, "diamic-tites" (unsorted piles of dirt and rock) deposited by gla-ciers in coastal seawater, "dropstones" in layered marine sediments that were transported by icebergs, and stream deposits produced by glacial meltwater. What is excep-tional about these glacial deposits is the evidence that they formed at low latitudes. Magnetic evidence for an equatorial paleolatitude comes from studies of the ori-entation of magnetic mineral grains, mostly magnetite, in sediments and volcanic rocks. The orientation of these grains is controlled by Earth's geomagnetic field, and thus is different in the tropics and high latitudes. The magnetic evidence involves carefully designed tests showing that the present orientation has been preserved since the time of deposition. Of particular interest are "fold tests" applied to sedimentary rocks judged to have been deposited as flat layers that were then deformed, or folded, shortly after their deposition. Samples of these rocks give a large range of paleolatitudes. However, the paleolatitudes converge to a single value when one cor-rects for the effects of folding and asks what the paleo-latitude would be if one "undeformed" the beds so that the strata were once again flat.

Evidence for global glaciation at 2.4 Ga and 1.9 Ga, the times of the two earliest events, is sparse At 2.4 Ga, low-latitude glaciation is recorded in the Transvaal Su-pergroup, central South Africa (Kirschvink et al. 2000); a stratigraphic section covering the relevant time period is shown in figure 3.1. The basal Koegas Formation con-tains sandstones and carbonates typical of low-latitude

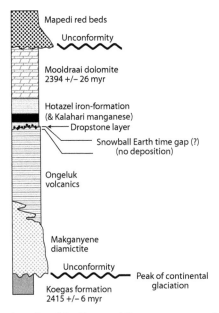

Fig. 3.1. Stratigraphy of the Transvaal Supergroup, central South Africa. From bottom to top: The Koegas is an equatorial limestone formation; the Makganyene diamictite is a glacial deposit; Ongeluk volcanics allow dating and the determination of an equatorial paleo-position (from paleomagnetic studies); the dropstone layer signifies that glaciers discharged into the oceans and melted; the Hotazel iron formation and Kalahari MnO_2 deposits signify the presence of some O_2 in air; and the Mooldraai Dolomite records typical tropical carbonate precipitation. From Kirschvink et al. (2000).

deposits. The overlying Makganyene diamictite contains glacially polished rocks and striated (stone-cut) surfaces among other features, reflecting its glacial origin (Evans et al. 1997). On top of the diamictite are the volcanic rocks of the Ongeluk Formation with magnetic properties indicating a tropical origin. These are overlain by the Hotazel Iron Formation, which contains dropstones at its base. There is evidence that other areas were glaciated at this time, including extensive deposits in the Huronian Supergroup of Ontario (Bekker and Kaufman 2007).

The King Leopold Sandstone in the Kimberly group, Western Australia, contains evidence for low-latitude glaciation at 1.8 Ga. There are pebble-cut striations indicating flow to the west. Then off in this direction, a pebble and boulder conglomerate is found, together with channels cut by meltwaters flowing under glaciers, all indicating glaciation. These deposits are overlain by marine sediments, indicating sea level glaciation. Paleomagnetic studies, including the fold test, puts the paleolatitude close to the equator (Schmidt and Williams 2008).

According to Allen and Etienne (2008), there were five distinct periods of glaciation between 770–630 Ma. The Sturtian, recently dated to 717 Ma (Macdonald et al. 2010), involved glaciation to sea level at the equator. Of all the Precambrian glacial events, the Marinoan is the best studied and most interesting. It is dated to about 635 Ma. Glacial rocks of this age are found on all continents. Evidence for glaciation includes observations similar to those discussed above: glacial diamictites and dropstones associated with marine sediments, striated rocks, and so on. The latitudinal distribution of these glacial sediments is known in two ways. The first is from paleomagnetic studies. The second is from the locations of the host formations, as inferred from reconstructions of past positions of continents and their orientation. In

total there is overwhelming evidence for widespread glaciation extending to sea level at low latitudes (Hoffman
and Li 2009).

The stratigraphic successions including the glaciation
typically have at their base limestones and siliceous sediments characteristic of tropical latitudes. These are then
overlain by glacial deposits, including diamictites with
striated rocks and dropstones. In the Elatina Formation,
Australia, the glacial strata include tidal "rhythmites" in-
dicative of deposition in a shallow marine environment.
Studies of magnetic properties in rhythmites established
that the glacial deposits span several reversals of Earth's
magnetic field, which puts their duration at a few million
years. Additional evidence for lengthy glaciation comes
from the iridium content of the glacial deposits. Iridium
in these deposits originates mainly from micrometeorites, whose flux to Earth's surface is fairly well known.
The time required to supply the iridium accumulation is
on the order of 3–12 Myr (Bodiselitsch et al. 2005). The
glacial deposits are in turn overlain by a "cap dolostone,"
a dolomite deposit that is typically about 10 m thick. The
cap dolostones may be overlain by very thick deposits of
$CaCO_3$ (Hoffman and Schrag 2002).

CAUSES OF SNOWBALL GLACIATIONS AND DEGLACIATIONS

Kirschvink (1992) offered a dramatic and parsimonious
explanation for these observations. Like others, he argued that if continents were glaciated in the tropics, they

would have been glaciated everywhere else too. With continents covered with ice to the tropics, much or all of the oceans would be frozen over as well, giving rise to the term "snowball earth." Earth would maintain itself in this condition, because the high albedo would stabilize the ice cover. Volcanoes would, however, continue to emit CO_2 to the oceans and atmosphere. Eventually, CO_2 would rise to a level high enough to melt ice in the tropical ocean. Local melting would lead to a lower albedo, local warming, more melting, an even lower albedo, and on and on. In this condition, the feedbacks would all be positive. That is to say, interactions between different parts of the climate system would all push in the same direction. Under these conditions, Earth would rapidly deglaciate.

A number of observations, mostly summarized by Hoffman and Schrag (2002), are consistent with this scenario, although not exclusively so. First, there is a large shift in the isotopic composition of carbon across the glaciation, to a value representative of the bulk Earth (about −5‰ [where 1‰ = 0.1%;]; box 1). This is the result one would expect if a large amount of CO_2 from Earth's interior accumulated in the atmosphere. Second, glacial deposits are overlain by the "cap dolostone," which contains various structures suggesting that it originated by rapid accumulation in storm-dominated waters (Hoffman and Schrag 2002). Rapid deposition could reflect a very large flux of $CaCO_3$ to the oceans caused by the rapid dissolution of continental rocks. This flux would in turn be the consequence of the high atmospheric CO_2

concentrations and very warm temperatures that would have prevailed with deglaciation. Third, after an absence from the sedimentary record lasting over one billion years, banded iron deposits are once again found, now in association with the snowball deposits. Banded iron formations are thought to form when the deep ocean is oxygen free, allowing dissolved Fe^{2+} to accumulate, while O_2 is produced by photosynthesis in the surface ocean. These might have been the conditions during snowball earth; in the deep ocean, respiration could consume dissolved O_2, making Fe^{2+} soluble and allowing its deep water concentration to rise. When deep water then mixed to the surface, Fe^{2+} would be rapidly oxidized, and precipitate.

Many of these views have been vigorously contested, but let us consider five points of widespread agreement. It is generally agreed that (1) glaciation did in fact extend to sea level in the tropics, (2) glaciation was a global event, (3) much of the ocean was ice covered, (4) deglaciation was rapid, and (5) there were large changes in the local or global carbon cycles as recorded by carbon isotope stratigraphy.

On the other hand, two critical points remain controversial. One concerns the extent of ice cover. The Kirschvink-Hoffman view is that the continents and oceans were completely covered in ice. An alternative is that there were open-ocean areas in the low latitudes ("slushball Earth"), a scenario that appears to be possible dynamically (Micheels and Montenari 2008). Since the floor of the open ocean is destroyed after about 100 Myr, when it is subducted back into Earth's interior, there is, alas, no sediment beyond the continental shelf or slope that might allow us to investigate the extent of ice cover. This limitation is problematic. The distinction between snowball and slushball is important because, in the latter case, the ocean and atmosphere communicate, allowing exchange of O_2 and CO_2 among other properties. The other, related, controversy concerns whether the Marinoan glaciation was a single, nearly static event, or whether there were repeated glaciations and an active hydrologic cycle. Because the surface was ice covered and very cold, one would expect slow rates of sublimation, little precipitation, and relatively little motion of the ice sheets. However, a number of authors have interpreted sediments from the Marinoan event as recording dynamic glaciers and repeated glaciations (Leather et al. 2002). Many specific points are also under discussion.

ENTERING INTO SNOWBALL GLACIATIONS

Recognizing that different snowballs may have had different causes, we can now step back and ask why Earth might have entered a snowball or slushball state in the first place. Explanations have been proposed that follow from our discussion of the faint young sun: the CO_2 concentration dropped to values too low to sustain a water-covered planet, high concentrations of reduced greenhouse gases (CH_4, C_2H_6, OCS) were oxidized too rapidly to allow CO_2 to rise in response, or the atmosphere somehow acquired a load of reflecting particles great enough to cool the planet and initiate glaciation. Part of the story appears to involve the presence or absence of O_2 in the atmosphere, and we digress to briefly discuss the origin and concentration history of this gas.

Earth is widely understood to have begun as a chemically reduced body. This view comes from the understanding that Earth accreted from a solar nebula dominated by hydrogen, and is supported by the presence of Earth's large metallic iron core. The atmosphere and surface rocks became oxidized through the agency of photosynthesis. In this process, plants split water into H_2 and O_2 and then use the H_2 to reduce CO_2 to organic matter.

Oxygen (O_2) is first a waste product that is not essential to the basic process of photosynthesis, and may in fact be harmful. It is not surprising, therefore, that photosynthesis can proceed using sources of H_2 other than H_2O. The most important alternative is sulfur photosynthesis, wherein plants split H_2S (hydrogen sulfide) into H_2 and S. There are actually two bacterial groups that carry out sulfur photosynthesis, the green sulfur bacteria and the purple sulfur bacteria. By 3.7 Ga, these two groups had developed different, albeit related, mechanisms ("photosystems") to extract H from H_2S. Sometime before 2.5 Ga, cyanobacteria appropriated both photosystems, combined them, and developed the capability of extracting H from H_2O, which is more difficult than extraction from H_2S. The ability to derive H atoms from ubiquitous water conferred a major advantage, and oxygenic photosynthesis eventually came to dominate the biosphere.

Its by-product, O_2, would oxygenate the atmosphere, oceans, and surficial rocks.

As explained in the previous chapter, oxygenation of the atmosphere is recorded by the changing pattern of sulfur isotope abundance in sedimentary compounds. There is a dramatic change at around 2.5 Ga, which is widely regarded as heralding levels of atmospheric O_2 high enough to form an ozone shield in the stratosphere (about 10 ppm). Kopp et al. (2005), among others, noted that the increase in atmospheric O_2 levels at 2.5 Ga corresponds to a time of snowball glaciation. They speculated that the appearance of O_2 in air would have quickly led to oxidation of reduced gases that may have been largely responsible for the Precambrian greenhouse (most likely CH_4, C_2H_6, and OCS). If the oxidative removal occurred over millions of years, CO_2 could have increased in response. However, more rapid removal would have lead to glaciation.

By 1.8 Ga, low levels of O_2 would have long been present in the atmosphere. We need to look for other mechanisms to explain the snowball glaciations of the Neoproterozoic (1 Ga–542 Ma). One hypothesis links cold climates to the meridional distribution of continents during the Neoproterozoic. As discussed in detail in chapter 4, CO_2 levels in air must rise until temperatures are high enough that consumption by weathering balances emissions. If continents are concentrated at low latitudes where temperatures are warmer than average, the equilibrium CO_2 concentration and global temperature would be much lower than if continents were concentrated at high latitudes (Schrag et al. 2002). During the Neoproterozoic, continents were centered at low latitudes, and weathering would have been rapid. Weathering thus may have drawn down CO_2 to a low enough value that high latitudes would become glaciated. Glaciers reflect sunlight and thereby cool adjacent regions. This cooling could lead to further growth of ice sheets, leading to lower albedo, then to cooling, and finally leading to more ice. Donnadieu et al. (2004) suggested an alternative in which it was the rifting, or breakup of the supercontinent Rodinia, that led to glaciation, at least for the Sturtian (the Neoproterozoic glaciation at ~717 Ma). The idea is that, after breakup, each chunk of the supercontinent would be surrounded by seas, rainfall would increase, and the weathering rate would rise. The presence of large areas of easily weathered basaltic rocks, produced by the breakup, would further accelerate weathering. In response, the CO_2 concentration would fall, again helping to induce glaciation.

Low-latitude continents would create a large area of shallow marine sediments in the tropics. Augmented by high runoff from tropical lands, this area would rapidly accumulate organic matter, again tending to lower CO_2 in the ocean and atmosphere. The buried organic matter might eventually be returned to the atmosphere as CH_4, which would strongly augment the CO_2 greenhouse. Temperatures would warm independent of CO_2, and CO_2 would fall to maintain the balance with emissions. In this scenario, Earth becomes dependent on methane to stay above freezing. This possibility would be an uncertain means of maintaining a warmer Earth, because CH_4 has a relatively short atmospheric lifetime and can diminish in concentration either because the source flux slows or because it is oxidized more rapidly. Eventually CH_4 would fail, Earth would cool in the face of low CO_2 concentrations, and a snowball or slushball would develop (Schrag et al. 2002).

Finally, Pavlov et al. (2005) proposed an extraterrestrial origin for global glaciation. At Earth's distance from the sun and beyond, interplanetary dust particles are normally deflected from the planet by the electromagnetic field of the heliosphere, which originates with protons, helium nuclei, and electrons from the sun's surface. Periodically, however, the solar system passes through a "giant molecular cloud" of dust and gas. The largest features, encountered every Gyr (one billion years) or so, collapse the heliosphere and allow dust to fall into Earth's atmosphere. There, it would cool the planet by reflecting the sun's light. According to Pavlov et al. (2005), passage through a giant molecular cloud with an H_2 density of 7000 atoms cm^{-3} would cause a temperature drop equivalent to that caused by dialing down solar luminosity by about 4%. The decrease in surface heating could initiate the cycles of cooling and ice sheet growth, leading to global glaciation.

REFERENCES

Papers with asterisks are suggested for further reading.

*Allen, P., and J. Etienne (2008), Sedimentary challenge to Snowball Earth, *Nature Geoscience*, *1*, 817–825.

Bekker, A., and A. Kaufman (2007), Oxidative forcing of global climate change: A biogeochemical record across the oldest Paleoproterozoic ice age in North America, *Earth and Planetary Science Letters*, *258*, 486–499.

Bodiselitsch, B., C. Koebert, S. Master, and W. Reimold (2005), Estimating duration and intensity of Neoproterozoic snowball galciations from Ir anomlies, *Science*, *308*, 239–242.

Donnadieu, Y., Y. Godderis, G. Ramstein, A. Nedelec, and J. Meert (2004), A "snowball Earth" climate triggered by continental break-up through changes in runoff, *Nature*, *428*, 303–306.

Evans, D., N. Beukes, and J. Kirschvink (1997), Low-latitude glaciation in the Palaeoproterozoic era, *Nature*, *386*, 262–266.

Grotzinger, J. (1986), Cyclicity and paleoenvironmental dynamics, Rocknest platform, northwest Canada, *Bulletin of the Geologic Society of America*, *97*, 1208–1231.

*Hoffman, P., and D. Schrag (2002), The snowball Earth hypothesis: Testing the limits of global change, *Terra Nova*, *14*, 129–155.

Hoffman, P., and Z.-X. Li (2009), A palaeogeographic context for Neoproterozoic glaciation, *Palaeogeography, Palaeoclimatology, Palaeoecology*, *277*, 158–172.

Hyde, W. T., T. J. Crowley, S. Baum, W. R. Peltier, (2000), Neoproterozoic "snowball Earth" simulations with a coupled climate/ice-sheet model, *Nature, 405*, 425–429.

*Kirschvink, J. (1992), *Late Proterozoic low-latitude global glaciation: The snowball Earth, In The Proterozoic Biosphere: A Multidisciplinary Study*, J. W. Schopf and C. Klein (eds.), Cambridge, U.K., Cambridge University Press.

Kirschvink, J., E. Gaidos, L. Bertani, N. Beukes, J. Gutzmer, L. Maepa, and R. Steinberger (2000), Paleoproterozoic snowball Earth: Extreme climatic and gochemical global change and its biological consequences, *Proceedings of the National Academy of Sciences of the United States*, *97*, 1400–1405.

Kopp, R., J. Kirschvink, I. Hilburn, and C. Nash (2005), The Paleoproterozoic snowball Earth: A climate disaster triggered by the evolution of oxygenic photosynthesis, *Proceedings of the National Academy of Sciences of the United States*, *102*, 11131–11136.

Leather, J., P. Allen, M. Brasier, and A. Cozzi (2002), Neoproterozoic snowball Earth under scrutiny: Evidence from the Fiq glaciation of Oman, *Geology*, *30*, 891–894.

Macdonald, F., M. Schmitz, J. Crowley, C. Roots, D. Jones, A. Maloof, J. Strauss, et al. (2010), Calibrating the Cryogenian, *Science*, *327*, 1241–1243.

Micheels, A., and M. Montenari (2008), A snowball Earth versus a slushball Earth: Results from Neoproterizoic climate modeling sensitivity experiments, *Geosphere*, *4*, 401–410.

Pavlov, A., O. Toon, A. Pavlov, and J. Bally (2005), Passing through a giant molecular cloud: "Snowball" glaciations produced by interstellar dust, *Geophysical Research Letters*, *32*. doi: 10.1029/2004GL028190.

Pierrehumbert, R. T., D. S. Abbot, A. Voigt, and D. Koll (2011), Climate of the Neoproterozoic, *Annual Reviews of Earth and Planetary Sciences*, *39*, 417–460.

Schmidt, P., and G. Williams (2008), Paleomagnetism of red beds from the Kimberley Group, Western Australia: Implications for the palaeogrography of the 1.8 Ga King Leopold glaciation, *Precambrian Research*, *167*, 267–280.

Schrag, D., R. Berner, P. Hoffman, and G. Halverson (2002), On the initiation of a snowball Earth, *Geochemistry, Geophysics, Geosystems, 3.* doi: 10.1029/2001GC000219.

Ueno, Y., M. Johnson, S. Danielache, C. Eskebjerg, A. Pandey, and N. Yoshida (2009), Geological sulfur isotopes indicate elevated OCS in the Archean atmosphere, solving faint young sun paradox, *Proceedings of the National Academy of Sciences of the United States, 106,* 14784–14789.

Wikinson, B., N. Diedrich, C. Drummond, and E. Rothman (1998), Michigan hockey, meteoric precipitation, and rhythmicity of accumulation on peritidal carbonate platforms, *Bulletin of the Geological Society of America, 110,* 1075–1093.

Yang, J., W. R. Peltier, and Y. Hu (2012), The initiation of modern "Soft Snowball" and "Hard Snowball" climates in CCSM3. Part I: The influences of solar luminosity, CO_2 concentration, and the sea ice/snow albedo parameterization, *Journal of Climate, 25,* 2711–2736.

4 REGULATION OF THE EARTH SYSTEM AND GLOBAL TEMPERATURE

SINCE LAND PLANTS EVOLVED ABOUT 400 MYR AGO, AND Earth then became biogeochemically modern, atmospheric CO_2 concentrations are thought to have varied by perhaps a factor of 10 or so. Concentrations fell to about 180 ppm during the recent ice ages. Most indicators suggest that they did not rise above about 1500 ppm during the last 400 Myr, although some point toward concentrations as high as 3000 ppm or more. For reference, modeling studies suggest that temperatures rise by about 2.5°C for every doubling of CO_2. Thus an increase of a factor of 8 in the CO_2 concentration (from 180 to 1440 ppm, for example) would cause average global temperatures to rise by about 8°C. Both past CO_2 concentrations and the climate sensitivity to a doubling of CO_2 are poorly known. Clearly, CO_2 variations are linked to dramatic climate variations, as we shall see. Nevertheless, it is interesting that during the Phanerozoic, CO_2 (and climate) variability has not been more extreme, as it was during the Neoproterozoic, for example. What mechanisms limit the magnitude of CO_2 variations, and keep the system in check?

In this chapter, we examine processes regulating CO_2 and global temperature. We focus on the leading paradigm; that the atmospheric CO$_2$ concentration and global temperature are fixed by feedback mechanisms balancing input and removal of CO$_2$. The geologic source of CO$_2$ to the atmosphere is outgassing from Earth's hot interior, and removal is through chemical reactions at Earth's surface ("weathering"). Our discussion deals with the Phanerozoic Eon. This period covers the last 543 Myr of Earth history. Plants and animals were present for most of this period, atmospheric O$_2$ was probably at about the present level, and Earth was in many ways geochemically modern.

BACKGROUND

A starting point for understanding processes regulating CO$_2$ is to look at the concentration of this gas in air. It is small: the preindustrial CO$_2$ concentration was only 280 ppm, and one might think that it could be easily changed. However, atmospheric CO$_2$ is in equilibrium with the oceans, which thereby provide a much larger buffering reservoir; the oceans contain 3200×10^{15} moles of CO$_2$ (actually dissolved inorganic carbon, which equals $CO_2 + HCO_3^- + CO_3^{2-}$) compared to 53×10^{15} moles CO$_2$ in preindustrial air. Changes in ocean circulation and the ocean carbon cycle will induce changes in atmospheric CO$_2$ by transferring this gas between the atmosphere and the very large deep ocean reservoir. Such changes led, for example, to glacial/interglacial CO$_2$ variations over the past million years. The timescale for these changes is several ocean mixing times, or thousands of years. Also relevant here is the land biosphere; there are about 200×10^{15} moles of carbon in living plants and soil. Changes in the extent of vegetation on land can lead to changes in atmospheric CO_2 levels that are significant, but not as large as atmospheric CO_2 changes induced by the oceans.

Glacial/interglacial CO_2 changes were modest in magnitude, ranging from about 180 to 300 ppm. Even this change corresponds to a ratio of 1.7 which, according to climate modeling studies, would lead to a deglacial warming of about 2°C. What processes might lead to larger CO_2 changes? To identify these processes we need to understand the dynamic nature of carbon in the oceans and atmosphere (fig. 4.1). Carbon dioxide is always being added to the oceans and atmosphere by "degassing" from Earth's interior. This process takes place in various ways; most obvious is the release of gases to air when volcanoes erupt. Another is as a consequence of the metamorphism of ocean sediments. In this process, biogenic $CaCO_3$, mostly from shells of microorganisms, reacts with biogenic SiO_2 fossils to release CO_2, which gradually leaks to the atmosphere:

$$CaCO_3 + SiO_2 \rightarrow CaSiO_3 + CO_2$$

When transferred to the atmosphere, CO_2 does not accumulate; it is continuously consumed to weather crystalline rocks at the surface. Weathering is the attack of CO_2 on minerals of igneous and metamorphic rocks brought to Earth's surface by volcanism and tectonics. Reactants include primary aluminosilicate minerals,

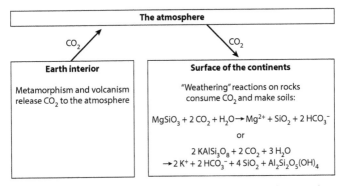

Fig. 4.1. An illustration of the geologic source and sink of CO$_2$ to the atmosphere. Carbon dioxide is transferred from Earth's interior to the atmosphere by volcanism and other processes, including metamorphism of sediments. It is consumed by weathering, represented here by simple reactions. Carbon dioxide rises to a level where global average temperature leads to consumption by weathering at a rate that balances input by volcanism.

CO$_2$, and water. Products include dissolved cations, dissolved bicarbonate (the neutralization product of the reaction), dissolved SiO$_2$, and residual insoluble minerals with lower concentrations of SiO$_2$ and cations than the reactant. A typical example is the weathering of potassium feldspar. This reaction consumes CO$_2$ and produces dissolved SiO$_2$, dissolved K$^+$, dissolved HCO$_3^-$, and insoluble kaolinite:

$$2KAlSi_3O_8 + 2\,CO_2 + 3\,H_2O \rightarrow Al_2Si_2O_5(OH)_4 + 4\,SiO_2 + 2\,K^+ + 2\,HCO_3^-.$$

Dissolved SiO$_2$, K$^+$, and HCO$_3^-$ are transported by rivers to the ocean. HCO$_3^-$ does not accumulate in the oceans;

it is continuously removed by $CaCO_3$ precipitation and other slower but critical processes that are not well understood.

Depending on how one does the calculation, the time required for volcanism to replace all the CO_2 in the surface reservoirs (atmosphere, ocean, and biosphere) is hundreds of thousands to millions of years. The longer timescale is relevant for us. It seems that, over this interval, the atmospheric CO_2 level is stabilized and regulated by a simple feedback mechanism: atmospheric CO_2 adjusts to the level where the temperature is such that weathering removes CO_2 as fast as it is supplied by volcanism. Walker et al. (1981) first described how this feedback would work. Weathering, like most chemical reactions, is faster at higher temperatures. If the atmospheric CO_2 concentration leads to warm temperatures and rapid CO_2 uptake that is greater than the input from Earth's interior, weathering is removing CO_2 faster than it is supplied by volcanism. Atmospheric CO_2 and global temperature will then fall toward the level where weathering equals supply. If the atmospheric CO_2 concentration is anomalously low, temperatures will be cold, and weathering will remove CO_2 slower than it is supplied. Carbon dioxide will then rise toward the level where Earth is warm enough so that removal balances supply. This paradigm invokes a simple feedback mechanism to balance the atmospheric CO_2 budget, and fix global temperature. At the same time, it offers a first-order explanation about why there have apparently been significant CO_2 variations: supply has changed through time, becoming faster when mountain building, seafloor spreading, and continental volcanism accelerate and cause faster release of CO$_2$ from Earth's interior to the atmosphere.

A MATHEMATICAL MODEL TO RECONSTRUCT THE ATMOSPHERIC CO$_2$ CONCENTRATION

In 1983, Robert Berner and colleagues developed a mathematical model of the Earth system describing the atmospheric CO$_2$ balance (Berner et al. 1983), and they elaborated on this model in subsequent papers (Geocarb Model: Berner 2006; Berner and Kothavala 2001; Berner 2009). There are two aspects to their models. The first describes Earth tectonics and simulates the rate, over time, at which CO$_2$ is transferred from the interior to the surface (supply side). The second describes the wide range of surface processes that influence CO$_2$ consumption by weathering (demand side).

The model recognizes that a number of processes transfer CO$_2$ from the interior to the atmosphere. It assumes that the overall transfer rate scales with the rate of seafloor spreading, which is related to underwater volcanism. This rate is not directly known for the past. However, it can be estimated from changes in sea level as inferred from the distribution of sediments on the continents (Gaffin 1987), because a higher sea level means more widespread deposition of ocean sediments on the continental landmass. There is a connection between sea level and spreading rate because when seafloor spreading is faster, the ocean rocks underlying the seawater are hotter, the ocean floor stands higher, sea level rises, and the oceans cover a greater extent of the continental land-mass. The bottom line: More continent covered by ocean sediments means faster sea floor spreading and more rapid degassing of CO_2.

Berner and colleagues invoked a broad spectrum of Earth surface processes that influence weathering rates. These include (1) weathering modeled as a process that breaks down calcium and magnesium silicates; (2) hydro-thermal processes at midocean ridges as a phenomenon that removes Mg^{2+} from seawater into rocks of the oce-anic crust, in exchange for Ca^{2+}; (3) the exchange of Ca for Mg when limestone ($CaCO_3$) is transformed into dolomite ($CaMg(CO_3)_2$); (4) the effects of organic car-bon burial (removes CO_2) and organic carbon weather-ing (i.e., the oxidation of organic carbon in sediments, which adds CO_2 to air); (5) the roles that sedimentation and weathering of FeS_2 and SO_4^{2-} play in the carbon cycle; (6) the estimation of precipitation over the different con-tinents from modeling studies, invoking the past posi-tions of the continents as inferred from paleomagnetic data; and (7) the estimation of paleoelevation and the dependence of weathering rate on altitude. Finally, these authors included the role of the land biosphere in weath-ering. Plants promote weathering by excreting organic acids into soils, and by injecting CO_2 derived from root respiration and rotting vegetation into the soils.

In their analysis, two salient influences lead to chang-ing CO_2 concentrations during the Phanerozoic, the last 543 Myr of Earth history (fig. 4.2). First, the model

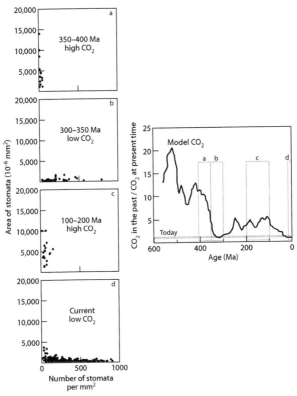

Fig. 4.2. Simulated values of the atmospheric CO$_2$ concentration during the Phanerozoic, and implications for the atmospheric CO$_2$ level from the size and abundance of stomata in fossil leaves. *Right*: Atmospheric CO$_2$ concentration versus time over the Phanerozoic, as simulated by the Geocarbsulf model of Berner (2006). The boxes labeled a–d mark the times corresponding to the panels on the *left*. *Other 4 panels*: Area of stomata versus number density of stomata for leaves from four time intervals. Large numbers of small stomata indicate adaptation to low CO$_2$; small numbers of large stomata indicate adaptation to high CO$_2$. The stomatal data and model generally agree on periods of high and low CO$_2$.

simulates high concentrations of CO_2 and warm temperatures at times in the past when sea level was high, and the inferred rate of seafloor spreading was high; these times were at about 542–350 Ma, and 100–200 Ma. Second, the model simulates disproportionately high concentrations of CO_2, and very warm temperatures, prior to the evolution of land plants at about 400 Ma. The reason is that, if plants are not facilitating weathering, atmospheric CO_2 needs to be much higher to promote weathering at a rate that balances input.

TESTING BERNER'S MODEL

Berner's model is likely to be qualitatively correct in assessing processes that play first and second order roles in regulating the atmospheric CO_2 burden, and the mechanisms through which these processes act. The model mathematically describes a whole range of processes that are extremely complex and are often not well understood in quantitative terms. Therefore it is appropriate to ask how seriously we should take the simulated CO_2 history. We adopt the view is that it is likely to qualitatively capture major CO_2 trends during the Phanerozoic, and therefore worthy of testing. In fact, there are a number of features of the CO_2 curve that are supported by proxy indicators of past CO_2 concentrations.

One line of support for the Berner curve comes from recent studies of the size and shape of stomata in fossil leaves. Stomata are the organs in plants that open to admit CO_2. Plants must open their stomata in order to photosynthesize, but they need to regulate this opening judiciously because it also leads to the loss of water. According to Franks and Beerling (2009), plants modify the number and size of their stomata in response to changes in the CO$_2$ concentration of air. When the CO$_2$ concentration is low, plants have a large number of small stomata. The advantage of small size is that, when stomata are shallow, CO$_2$ in air only needs to diffuse a small distance to enter the leaf. When the CO$_2$ concentration is high, plants make large, deep stomata; CO$_2$ needs to diffuse further, but this is not a problem when CO$_2$ is abundant in air.

The broad patterns of stomatal size and number support the general pattern of CO$_2$ variations suggested by Berner's Geocarb III model (fig. 4.2). In this figure, the modeled CO$_2$ curve is shown in the upper right panel. In the other panels, stomatal area is plotted versus stomatal number density for times of particular CO$_2$ concentrations. In modern leaves, stomatal size is low and the number density is high, consistent with expectations. Between 100–200 and 350–400 Ma, Geocarb III predicts high CO$_2$ concentrations; stomatal sizes are large, and numbers are small. Finally, between 300–350 Ma, Geocarb III predicts low CO$_2$ concentrations; small stomatal sizes and high number densities support this prediction.

More support for the model CO$_2$ curve comes from paleosol carbonates. These are CaCO$_3$ rocks forming within soils. The δ^{13}C of soil CO$_2$ is affected by soil respiration (Ekart et al. 1991). (For δ^{13}C, or the ratio of ^{13}C to ^{12}C, see box 1.) As organic matter decomposes in soils, releasing metabolic CO_2, the $\delta^{13}C$ of soil CO_2 changes from the atmospheric value to the organic matter value. It turns out that the atmospheric value back through time is rather well known; limestones preserve a record of the $\delta^{13}C$ of dissolved inorganic carbon in surface seawater, to which the atmospheric value is related by chemical and isotopic equilibrium. Through nearly all of the Phanerozoic, the $\delta^{13}C$ of atmospheric CO_2 was ~ −7 ‰. Plants, which discriminate against ^{13}C by about −20‰, have a $\delta^{13}C$ of about −27‰. The CO_2 of gases (and solutions) in soils will be a mixture of two components: atmospheric CO_2 with $\delta^{13}C$ ~ −7‰, and metabolic CO_2 derived from rotting plant debris, with $\delta^{13}C$ ~ −27‰. If atmospheric CO_2 is high, it will contribute more to the mix, and $\delta^{13}C$ of paleosol carbonates will be isotopically heavy (more like air). If atmospheric CO_2 is low, metabolic CO_2 will dominate the mix in soil air, and paleosol carbonates will be isotopically light (more like vegetation).

Ekart et al. (1991) applied this approach to a large suite of paleosols. Various confounding processes introduce a lot of noise into the record. Nevertheless, key features agree with the model of Berner. According to the paleosol index, CO_2 was low from about 380 to 300 Ma, whereas it was low from about 350 to 260 Ma in Berner's model. Carbon dioxide was elevated from 230 to 65 Ma in both the model and paleosol data. The paleosol record shows a sharp CO_2 minimum at about 65 Ma, which is absent from the model. In both the model and the paleosol reconstruction, CO_2 decreased over the past 55 Ma.

Studies of $\delta^{13}C$ of organic carbon, in both fossil leaves and marine sediments, also give some support to the Berner curve. In both cases, the tracer quality of $\delta^{13}C$ data rests on the fact that plants discriminate against the heavy carbon isotope, ^{13}C, during photosynthesis. Today, this discrimination causes the $\delta^{13}C$ of plants to be $\sim -27‰$, as stated above. If the CO$_2$ concentration is high, photosynthesis "selects" from a large carbon pool and can exercise its preference for ^{12}C. In this case, $\delta^{13}C$ will be $< -27‰$. If the CO$_2$ concentration is low, CO$_2$ is in short supply, and photosynthesis must use more of the available CO$_2$. Hence, $\delta^{13}C$ of organic carbon is isotopically light when CO$_2$ is high, and heavy when CO$_2$ is low. Fletcher et al. (2007) analyzed 11 fossil leaf samples ranging from 170 to 60 Ma. The $\delta^{13}C$ values were systematically lighter than in modern plants, and the aggregate data suggest an atmospheric CO$_2$ concentration about three times the present level. This result agrees well with Berner's simulation for the same interval.

Finally, Pagani et al. (2005) summarized a large data set of $\delta^{13}C$ analyses on marine organic carbon. Studies on bulk organic carbon in ocean sediments would be perilous because different samples would have different mixes of organic chemicals, each of which might be isotopically fractionated to a different degree. They therefore restricted their study to alkenones, a single type of organic molecule made by a small number of organisms that is very resistant to degradation. Through a number of assumptions, they were able to convert their $\delta^{13}C$ measurements into paleoatmospheric CO$_2$ values, albeit with

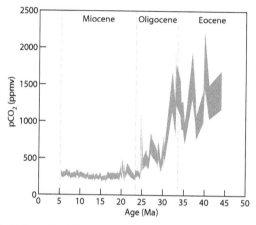

Fig. 4.3. Carbon dioxide versus time as inferred from the carbon isotope composition of alkenones during the Cenozoic (Pagani et al. 2005). Lower values of $\delta^{13}C$ indicate more selectivity for the light isotope (^{12}C) and, hence, higher CO_2 in the past.

a large uncertainty. Their results are shown in figure 4.3; the record stops at about 50 Ma because even alkenones are degraded in older sediments. Pagani et al.'s record is similar to Berner's in showing a CO_2 decrease of about a factor of four during the Cenozoic. However, it differs from the Geocarb III model in that CO_2 drops to the present level at about 30 Ma in the data, compared to about 10 Ma in the model.

In summary, Berner's model simulates very high CO_2 concentrations prior to 350 Ma, low values from about 350 to 260 Ma, high but variable values between 170–60 Ma, and decreasing levels over the past 55 Ma. Proxy studies generally support this pattern. Therefore the model appears to give an accurate first-order picture of CO$_2$ variations during the Phanerozoic, and is useful in this context. However, the model is highly uncertain in terms of magnitudes, and the timing of CO$_2$ changes will likely require modifications.

CARBON DIOXIDE CHANGES IN THE CONTEXT OF LARGE-SCALE CHANGES IN THE EARTH SYSTEM DURING THE PHANEROZOIC

Changes in atmospheric CO$_2$ concentrations are part of a broader cycle of Earth's surface changes during the Phanerozoic; this is referred to as the oscillation between "calcite seas" and "aragonite seas" (fig. 4.4). Calcite seas correspond to a period of warm climate, high atmospheric CO$_2$, low seawater Mg^{2+}/Ca^{2+}, precipitation of CaCO$_3$ cements as the calcite polymorph (mineral form), and high sea levels leading to submergence of large continental areas ("high onlap"). Aragonite seas correspond to a cool climate and glaciation, low CO$_2$, high seawater Mg^{2+}/Ca^{2+}, precipitation of aragonite cements, and continents standing entirely proud of the ocean ("high offlap"). These changes form the context in which we can place the specific climate changes to be discussed in the rest of this book.

Evidence for past seawater composition comes from two sources. The first is a series of classic papers on the chemical composition of brine inclusions in evaporites (Horita et al. 2001; Lowenstein and Timofeeff 2008; Lowenstein et al. 2001; Timofeeff et al. 2006). Marine

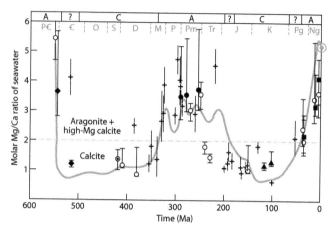

Fig. 4.4. Basic data describing calcite and aragonite seas (Stanley et al. 2010). The gray line is from the model of Demicco et al. (2005). The "+" represents the ratio of Mg/Ca in seawater inferred from studies of fossil $CaCO_3$ shells. Filled and open circles, diamonds, and triangles represent seawater Mg/Ca ratios inferred from brine inclusions in evaporites. In the bar at the top, "A" refers to aragonite seas and "C" refers to calcite seas. The horizontal line at a Mg/Ca ratio of 2 is the putative boundary between calcite and aragonite seas. The modern seawater Mg^{2+}/Ca^{2+} ratio is 5.2. See Stanley et al. for additional citations to the plotted data.

evaporites form from the evaporation of seawater in isolated basins, and brine inclusions are droplets of evaporated water captured within the depositing salts. During evaporation, the composition of the inclusions follows a trajectory that reflects, and can be used to reconstruct, the starting composition of unevaporated seawater. The second archive of seawater chemistry is the study of the trace element composition of unaltered

CaCO$_3$ skeletons of fossil echinoderms (Dickson 2002, 2004). The skeletal Mg/Ca ratio of some groups of animals depends on the seawater ratio. There is also a relationship between seawater composition and the mineralogy of cements. The ion Mg^{2+} inhibits precipitation of calcite. A low ratio of Mg^{2+}/Ca^{2+} in seawater enables CaCO$_3$ to precipitate as calcite, the stable polymorph. A high Mg^{2+}/Ca^{2+} ratio leads to precipitation of metastable aragonite.

There is a strong empirical link between the Mg^{2+}/Ca^{2+} ratio of seawater and the CO$_2$ concentration of air. However, the mechanism remains to be understood. The atmospheric CO$_2$ concentration is regulated by input to the atmosphere from degassing, and removal by weathering. The ocean does not necessarily have a role to play in either of these processes. The seawater Mg^{2+}/Ca^{2+} ratio is regulated primarily by the removal rate of Mg^{2+}. Two main processes remove Mg: hydrothermal reactions between seawater and the volcanic bedrock of the oceans, and the precipitation of dolomite (CaMg(CO$_3$)) at the expense of CaCO$_3$. When these processes are rapid, [Mg^{2+}] will be low in seawater, and vice versa. Onlap heralds rapid Mg^{2+} removal, both because shallow seas favor dolomite deposition and because a high sea level is the consequence of rapid seafloor spreading and hence a faster removal of Mg^{2+} by seafloor reactions. A low seawater Mg^{2+}/Ca^{2+} ratio would be linked to higher pCO$_2$ if outgassing is faster when seafloor spreading accelerates. The connection between spreading rate and outgassing is controversial (Berner 2006; Coggon et al. 2010; Hardie

1996; Holland 2005), and at this time we do not completely understand it.

REFERENCES

Papers with asterisks are suggested for further reading.

Berner, R. (2006), Geocarbsulf: A combined model for Phanerozoic atmospheric O_2 and CO_2, *Geochimica et Cosmochimica Acta*, *70*, 5653–5664.

Berner, R., and Z. Kothavala (2001), Geocarb III: A revised model of atmospheric CO_2 over Phanerozoic time, *American Journal of Science*, *301*, 182–204.

Berner, R., A. Lasaga, and R. Garrels (1983), The carbonate-silicate geochemical cycle and its effects on atmospheric carbon-dioxide over the past 100 million years, *American Journal of Science*, *260*, 641–683.

Berner, R. (2009), Phanerozoic atmospheric oxygen: New results using the Geocarbsulf model, *American Journal of Science*, *309*, 603–606.

Coggon, R. M., D.A.H. Teagle, C. E. Smith-Duque, J. C. Alt, and M. J. Cooper (2010), Reconstructing past seawater Mg/Ca and Sr/Ca from mid-ocean ridge flank calcium carbonate veins, *Science*, *327*, 1114–1117.

Demicco, R. V., T. Lowenstein, L. A. Hardie, and R. J. Spenser (2005), Model of seawater composition for the Phanerozoic, *Geology*, *33*, 877–880.

Dickson, J.A.D. (2002), Fossil echinoderms as monitor of the Mg/Ca ratio of Phaneorzoic Oceans, *Science*, *298*, 1222–1224.

Dickson, J.A.D. (2004), Echinoderm skeletal preservation: Calcite-aragonite seas and the Mg/Ca ratio of Phanerozoic oceans, *Journal of Sedimentary research*, *74*, 355–365.

Ekart, D., T. Cerling, I. Montanez, and N. Tabor (1991), A 400 million year carbon isotope record of depogenic carbonate: Implications for paleoatmospheric carbon dioxide, *American Journal of Science*, *299*, 805–827.

Fletcher, B., S. Brentnall, C. Anderson, R. Berner, and D. Beerling (2007), Atmospheric carbon dioxide linked with Mesozoic and Early Cenozoic climate change, *Nature Geosciences*, *1*, 43–48.

Franks, P., and D. Beerling (2009), CO$_2$-forced evolution of plant gas exchange capacity and water use efficiency over the Cenozoic, *Geobiology*, *7*, 227–236.

Gaffin, S. (1987), Ridge volume dependence on seafloor generation rate and inversion using long term sealevel change, *American Journal of Science*, *287*, 596–611.

Hardie, L. A. (1996), Secular variation in seawater chemistry: An explanation for the coupled secular variation in the mineralogies of marine limestones and potash evaporites over the past 600 Myr, *Geology*, *24*, 279–283.

Holland, H. D. (2005), Sea level, sediments and the composition of seawater, *American Journal of Science*, *305*, 220–239.

Horita, J., H. Zimmermann, and H. D. Holland (2001), Chemical evolution of seawater during the Phanerozoic: Implications from the record of marine evaporites, *Geochmica et Cosmochimica Acta*, *66*, 3733–3756.

Lowenstein, T. K., and M. N. Timofeeff (2008), Secular variations in seawater chemistry as a control on the chemistry of basinal brines: Test of the hypothesis, *Geofluids*, *8*, 77–92.

....

*Lowenstein, T. K., M. N. Timofeef, S. T. Brennan, L. A. Hardie, and R. V. Demicco (2001), Oscillations in Phanerozoic seawater chemistry: Evidence from fluid inclusions, *Science*, *294*, 1086–1088.

*Pagani, M., J. Zachos, K. Freeman, B. Tipple, and S. Bohaty (2005), Marked decline in atmospheric carbon dioxide concentrations during the Paleogene, *Science*, *309*, 600–603.

Stanley, S. M., J. B. Ries, and L. A. Hardie (2010), Increased production of calcite and slower growth for the major sediment-producing alga *Halimeda* as the Mg/Ca ratio of seawater is lowered to a "Calcite Sea" level, *Journal of Sedimentary Research*, *80*, 6–16.

Timofeeff, M. N., T. K. Lowenstein, M. A. Martins da Silva, and N. B. Harris (2006), Secular variation in the major-ion chemistry of seawater: Evidence from fluid inclusions in Cretaceous halites, *Geochimica et Cosmochimica Acta*, *70*, 1977–1994.

*Walker, J., P. Hays, and J. Kasting (1981), A negative feedback mechanism for the long-term stabilization of Earth's surface temperature, *Journal of Geophysical Research–Atmospheres*, *86*, 9776–9782.

5 THE LATE PALEOZOIC ICE AGES

EARTH WAS GLACIATED DURING MUCH OR MOST OF AN interval stretching from the Late Devonian to the Middle Permian, from about 370 to 260 Ma. Evidence for glaciation comes from two sources. First, there are "near-field" deposits on all continents that directly reflect glacial activity. Second, there are contemporaneous deposits in regions of low latitude that reflect changes in sea level associated with growth and decay of the ice sheets (sea level drops as ice sheets grow). Sea level would have changed over glacial-interglacial cycles as water was removed from the oceans to form ice sheets, then returned as the ice sheets melted. The amplitude of these changes in sea level is not well known. It may at times have been 80 m or more, which is about 60% of that associated with glacial-interglacial cycles in the Pleistocene. For reference, if the global land area is half that of the oceans, a change of 80 m corresponds to about 160 m of glacial ice spread uniformly over the surface of the continents. So the Late Paleozoic ice ages, judged from their areal extent, temporal span, and amplitude, were a big deal.

There is evidence, from observations of geologic deposits and from modeling, that atmospheric CO_2 concentrations were low during the Late Paleozoic ice ages, and this may have caused or contributed to the event. These ice ages correspond to a time when most landmasses were agglomerated into a single supercontinent over the South Pole, providing a favorable environment for the development of large ice sheets. There is emerging evidence that glacial cycles were, as in later times, regulated by changes in Earth's orbit around the sun.

In this chapter, we summarize physical evidence for ice ages during the Late Paleozoic, first from glacial deposits, and second from sedimentary cycles in unglaciated regions. We then look at syntheses examining changes in the extent of glaciation in time and space. Next, we summarize factors that may have made glaciation more likely in the Late Paleozoic, as well as the causes of glacial cycles. Finally, we briefly outline some major unresolved questions.

THE RECORD OF GLACIATION

"Near-field" evidence for glaciation comes from deposits or features directly linked to the activity of glaciers. One compelling indication for glaciation is the presence of polished basement surfaces that have striations formed by rocks dragged along the bottom of glaciers. These features are surprisingly common; an impressive example comes from the Late Devonian of the Parnaiba Basin, Brazil (Caputo et al. 2008). (Papers cited in this paragraph have pictures of the relevant features.) Also common in many locations are cobbles that have striated surfaces (e.g., Brezinski et al. 2008; Isaacson et al. 2008) (fig. 5.1). Another line of evidence is the presence of glacial

Fig. 5.1. (a) Late Devonian boulder with glacial striations in a diamictite, at Pisac, Peru. Isaacson et al. 2008. Late Devonian—earliest Mississippian glaciation in Gondwanaland and its biogeographic consequences. Palaeogeography, Palaeoclimatology, Palaeoecology, 268, no. 3-4, 126-142. © 2008 Elsevier. Adapted and reprinted with permission. (b) Striated clasts, with arrows showing direction of ice flow, from Lower Carboniferous, Pennsylvania/Maryland. Brezinski et al. 2008. Late Devonian glacial deposits from the eastern United States signal an end of the mid-Paleozoic warm period. Palaeogeography, Palaeoclimatology, Palaeoecology 268, no. 3-4, 143-151. © 2008 Elsevier. Adapted and reprinted with permission.

(c) Late Devonian pavement, glacially polished and striated, Eastern Brazil. Caputo et al. 2008. Late Devonian and Early Carboniferous glacial records of South America. The Geological Society of America Special Paper, 441, 161-173. © 2008 The Geological Society of America. Adapted and reprinted with permission.

(d) Dropstones in carved sediments of a proglacial lake, Late Devonian, Pennsylvania. Brezinski et al. 2008. Late Devonian glacial deposits from the eastern United States signal an end of the mid-Paleozoic warm period. Palaeogeography, Palaeoclimatology, Palaeoecology, 268, no. 3-4, 143-151. © 2008 Elsevier. Adapted and reprinted with permission.

rhythmites. Rhythmites are sediments with repeating lithology that can originate in a number of ways. The type of greatest interest to us derives from glacial outwash. In a proglacial lake (out in front of the melting ice sheet), there will be continuous sedimentation of fine grained material. However sedimentation of coarse grained material will be concentrated during springtime, when melting and runoff are greatest. The result is layered sediment with coarse-grained springtime layers, and fine-grained layers representing deposition during the rest of the year. Glaciers calving directly into the lake will melt and release the debris they bear. Dropstones will settle to the bottom, deforming the fine strata of rhythmites, and the rocks will eventually be draped by newly deposited sediments.

Perhaps the most common glacier sediments are diamictites. As noted earlier, these piles of rocks and dirt can have various origins. In Late Paleozoic strata, their glacial pedigree is established by the presence of faceted (cut) and striated boulders (Mory et al. 2008), by the presence of dropstones (Chakraborty and Ghosh 2008; Martin et al. 2008), and by proximity to glacial rhythmites or a striated basement (basement rocks polished by the flow of the glacier, with grooves cut by rocks dragged over the bed).

In the early Carboniferous, around the beginning of Late Paleozoic glaciation, South America, Africa, Arabia, India, Antarctica, and Australia were joined (fig. 5.2; (Frank et al. 2008). They were centered on the South Pole and formed the supercontinent of Gondwanaland. It was this supercontinent that was most extensively glaciated, primarily because other landmasses were concentrated in tropical and temperate latitudes. Glacial deposits on Gondwanaland date to different times. Figure 5.2 divides glaciation into three episodes, spanning times from 360 to 345 Ma, 325 to 310 Ma, and 300 to 290 Ma (Glacials I, II, and III are recorded by black, dark gray, and light gray areas, respectively). During the time of the Late Paleozoic glaciations, Gondwanaland continuously drifted over the South Pole. At 360 Ma, the Pole lay under what is now equatorial Africa and South America. Gondwana eventually moved so that central Africa, central Antarctica, and eastern Australia progressively crossed over the Pole. Glaciation occurred on lands far enough south that ice would be prevalent at low elevations. Land masses were thus glaciated sequentially as they slid over the Pole (or came close enough to sustain ice sheets).

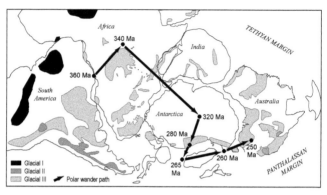

Fig. 5.2. Positions of Africa, South America, India, Antarctica, and Australia around 300 Ma. Black, dark gray, and light gray areas show the past locations of glacial deposits from three different periods (Glacial I, 365–350 Ma; Glacial II, 325–310 Ma; Glacial III, 300–285 Ma). The solid black line shows the positions of the continents as a function of time, illustrated in terms of the path of the South Pole as it passed over fixed continental positions. Ages indicate the time when the pole was located at the marked position between 360–250 Ma (Frank et al. 2008).

The earliest area to be glaciated was what is now tropical South America. Thus, the earliest evidence for Late Paleozoic glaciation comes from Brazil, Bolivia, and Peru (Caputo et al. 2008; Frank et al. 2008; Isaacson et al. 2008; see fig. 5.2), and dates to about 360 Ma. Later glaciations affected more southerly regions of South America, as well as Australia and Antarctica. Glaciation was most extensive between about 290–315 Ma, as inferred from the distribution of near-field deposits (Frank et al. 2008) and evidence of sea level change in low latitudes (Rygel et al. 2008). The last area to be "permanently" deglaciated was eastern Australia, which was near the South Pole at the end of the Permian (fig. 5.2). Glaciation here ended at about 260 Ma (Birgenheier et al. 2010; Fielding, Frank, and Isbell 2008; Fielding, Frank, Birgenheier, et al. 2008).

Finally, there is some evidence for Northern Hemisphere glaciation during the Late Paleozoic. Brezinski et al. (2008) report that Late Devonian glacial deposits extend from northeastern Pennsylvania into the northern region of Virginia and West Virginia. This glaciation appears to have been contemporaneous with the earliest Late Paleozoic glaciation in tropical South America, and thus records an interhemispheric event. It is particularly curious because Pennsylvania was at this time located in the warmer temperate latitudes. Siberia was later glaciated in the Middle Permian, at about 270 Ma and coincident with the late glaciation of eastern Australia (Fielding, Frank, and Isbell 2008, and citations therein). Siberia was located in the high northern latitudes and its glaciation is therefore unsurprising.

PALEOZOIC CYCLOTHEMS: THE RECORD OF GLACIOEUSTATIC SEA LEVEL VARIATIONS

The waxing and waning of the ice sheets is recorded in sedimentary successions called cyclothems (see also chapter 3). Cyclothems occur because nearshore sediments are very sensitive to local sea level. Global sea level falls as global ice volume grows, and rises as ice melts. In areas near large ice sheets, the relationship between local sea level and global ice volume is complicated by "isostatic effects," as the local crust is depressed by the weight of the glaciers. Sea level near ice sheets also reflects self-gravitation, wherein glaciers modify the elevation of the local ocean surface by gravitationally drawing seawater to them. One therefore accesses the sedimentary record of ice volume in tropical areas, where the effects of isostacy and self gravitation are small. In these areas, local sea level changes are closely linked to global sea level changes associated with the growth and decay of ice sheets.

In one idealized version of a cyclothem, the base of the unit, corresponding to interglacial times of highest sea level, is a black shale. This fine-grained sediment derives its black color from the presence of pyrite, which is linked to low-O_2 bottom waters and the decay of organic carbon in sediments. Low O_2 signifies a depth below a density change in the overlying water, which prevents O_2-rich surface waters from mixing. Fine-grained shales reflect sedimentation below the depth of vigorous wave action.

Further up the column, sediment types change in a way that indicates a rising sea level and, consequently, diminishing continental ice sheets. Black shale is replaced by gray shale, which is less extensively reduced (lower pyrite concentration). Next come deposits of $CaCO_3$, mostly in the form of cemented, coarse-grained skeletal debris associated with a progressively shallower water depth. These sediments can be overlain by beach sands, signifying deposition at sea level, followed by paleosols—ancient soils that must form above sea level. Deposits forming above sea level often include coals that formed in coastal wetlands; in fact, Late Paleozoic cyclothems host the most economically important coal beds in Europe, North America, and Australia. In one view, coals are associated with rising sea levels attending deglaciation, while some paleosols form in seasonally dry areas associated with maximum glaciation (Falcon-Lang et al. 2009). Further up the sediment column one finds sediment types that signify progressively deeper water depths as the glaciers melt, sea level rises, and the movie runs in reverse. Eventually one again encounters a black shale, indicating the highest sea level and maximum deglaciation. Then the cycle repeats.

The reader should be aware that this view of cyclothems is somewhat controversial. Sedimentary successions in cyclothems are far more complex than the idealized version described above. This complexity is partly due to changes in the amplitude of sea level variations, such that different groupings of sediment types will be present. There may also be local changes of sea level due to tectonic uplift or subsidence. Wilkinson et al. (2003) have shown that the succession of sediment types in Pennsylvanian-age cyclothems of Illinois is indistinguishable from a random sequence. They argued that cyclothems reflect changes in local sea level associated with random changes in sediment sources and the migration of rivers and beaches. This point is also discussed in the chapter on Precambrian glaciations.

We however favor the general consensus that most cyclothems reflect glacioeustatic changes in sea level. First, many of the variations in sediment types suggest large changes in local sea level—50 m or (much) more (Rygel et al. 2008). Second, the nature of cyclothems changes in the expected way when moving from sites near a continent to offshore areas of deeper seafloors (e.g., Heckel 1986). Third, one commonly finds that marine strata in cyclothems are eroded by paleovalleys with elevations of tens of meters. These valleys formed when sea level was low, exposing sediments and enabling erosion (e.g., Falcon-Lang et al. 2009; Hampson et al. 1999). Fourth, our detailed knowledge of the Late Pleistocene shows that the amplitude of sea level changes can vary dramatically from cycle to cycle, in which case one would not observe close repetition. Finally, the link between the Late Paleozoic glacial deposits at high latitudes and the presence of cyclothems at low latitudes is, to say the least, highly suggestive.

GLOBAL CLIMATE VARIATIONS
DURING THE LATE PALEOZOIC

In recent years, there have been several major technical advances leading to improvements in our understanding of Late Paleozoic glaciation. These in turn form the basis for a provisional narrative of global climate change during this interval. As already discussed, glaciation commenced in the Late Devonian and occured, discontinuously, until the Late Permian. The most intense glaciation was generally on the southerly areas of Gondwanaland. Glaciation leads to eustatic sea level change recorded by cyclothems dating to different intervals in different regions. Cyclothems of the Donets Basin, Russian Platform, date from about 330 Ma (Late Mississippian) to 295 Ma (Early Permian) (Davydov et al. 2010). Cyclothems from Arrow Canyon, Nevada, probably date from about 330 Ma to 300 Ma (Bishop et al. 2010). Cyclothems from the American midcontinent, which are very well developed in Kansas and Oklahoma, span the period from about 310 to 300 Ma (Heckel 2008).

According to the synthesis of Rygel et al. (2008), cyclothems are most extensive between about 335–300 Ma, but Late Paleozoic cyclothems extend to times as old as 350 Ma and as young as 260 Ma (fig. 5.3). The sedimentary evidence puts the greatest amplitude in sea level variations at about 300 Ma, corresponding to the Pennsylvanian/Permian boundary (Bishop et al. 2010; Fielding, Frank, and Isbell 2008). According to Bishop et al., the amplitude of sea level changes is generally greater between about 310–320 Ma than during the 10 Myr periods before and after.

The amplitude of sea level changes remains controversial. The basic problem is that we cannot accurately estimate water depth or elevation above sea level associated with the sediment types reflecting maximum submergence or emergence. For example, a black mud generally signals deeper water than limestone, but precise water depths can't be assigned to either of these facies.

According to Rygel et al. (2008), estimates of sea level variations based on the sediment types found in cyclothems range in amplitude from 10 to 100 m or more. Two recent papers use the oxygen isotope composition

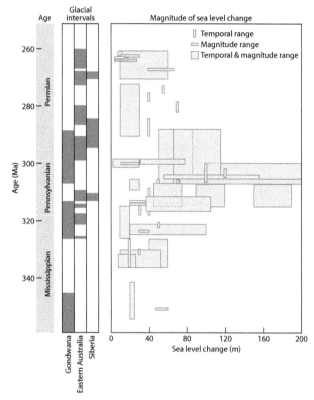

Fig. 5.3. Estimates, based on data from cyclothems, of sea level change as a function of time (Rygel et al. 2008). The panel on the left shows time, broken down into different geological intervals. The center plot ("Glacial intervals") uses dark bands to represent periods of time over which glacial deposits have been found in Gondwana-land (*left column*), Eastern Australia (*central column*), and Siberia (*right column*). In the panel on the right ("Magnitude of sea level change"), each box indicates a cyclothem sequence. The time interval is the period spanned by that sequence, and the distance interval is the maximum and minimum sea level change implied by the cyclothems in the sequence.

of phosphate in fossil conodonts to estimate glacioeustatic variations of sea level of about 100 m or more during the Middle–Late Pennsylvanian. (Oxygen isotopes in the PO_4^{3-} of conodonts record water temperature and ice volume changes as do O isotopes in the $CaCO_3$ of Foraminifera, the amoeboid protists whose skeletons are ubiquitous on the modern ocean floor. In box 1, we discuss the nature of stable isotope variations. In box 2, we discuss how this proxy works for $CaCO_3$; it works in a similar manner for phosphates.) On the other hand, Bishop et al. (2010) argue for amplitudes of 50 m or less based on several lines of evidence, including reevaluations of amplitudes implied by canyon cutting, and the dynamic nature of the cyclothem record. They suggest that multiple glacial centers contributed to Late Paleozoic glaciation, distinct from the Pleistocene pattern of great ice sheets growing from a small number of regions.

DYNAMICS OF THE LATE PALEOZOIC GLACIATIONS

According to the prevailing view, there were two causes of widespread glaciation during the Late Paleozoic. First, CO_2 was low (see chapter 4 and the summary below), predisposing Earth to glaciation. Second, there was a massive continental landmass at high southern latitudes that allowed nucleation and growth of large ice sheets. These ice sheets waxed and waned within longer periods of more or less continuous glaciation. Climate cycles of tens or hundreds of thousands of years have

Box 2
Oxygen Isotope Paleothermometry Using
Foraminifera and Other CaCO$_3$ Fossils

For the Cenozoic and beyond, the most important source of
paleoclimate information comes from the δ^{18}O of fossil Foraminifera. Foraminifera are amoeboid protists that secrete
shells of CaCO$_3$ (among other materials). "Planktonic Foraminifera" are ubiquitous in the upper levels of the oceans, and
are common in ocean sediments from, roughly, the shallower
half of the sea floor (CaCO$_3$ dissolves in the deeper, colder,
higher pressure environments). Benthic Foraminifera live on
the sea floor; they are much less abundant than planktonic
Foraminifera but are still always found in shallower sediments.
Studies of the oxygen isotope composition of these protists
play a central role in paleoclimate for two reasons. First, they
illuminate critical properties of past climates: ice volume, sea
surface temperature, and deep ocean temperature. Second,
they are common in CaCO$_3$ sediments, enabling paleoclimate
scientists to construct continuous, high-resolution climate records pertaining to the sea surface and the sea floor over long
periods of time.

Three factors determine the δ^{18}O of Foraminifera. The first
is the δ^{18}O of ambient seawater. The δ^{18}O of biogenic CaCO$_3$
directly tracks the δ^{18}O of the seawater from which the solid
phase is precipitated. The δ^{18}O of local seawater depends on
the average δ^{18}O of the global ocean. And the δ^{18}O of the global
ocean increases when ice sheets grow, because waters low in
^{18}O evaporate preferentially to be incorporated into continental ice sheets. The δ^{18}O of planktonic forams will change
through time because of both local salinity and global ice volume. For benthic Foraminifera, salinity changes are generally
less important than global ice volume.

The second factor determining the $\delta^{18}O$ of foraminiferal $CaCO_3$ is temperature. At Earth surface temperatures, ^{18}O is enriched in CO_3^{2-} with respect to the water with which CO_3^{2-} is in equilibrium. At very high temperatures, however, isotopes are completely scrambled between H_2O and CO_3^{2-}. Thus, the $\delta^{18}O$ of $CaCO_3$ must decrease as temperature rises.

The third factor determining the $\delta^{18}O$ of foraminiferal $CaCO_3$ is the "vital effect." Some foram species precipitate $CaCO_3$ out of isotopic equilibrium with water. The solution to this confounding phenomenon is either to exclude these taxa, or document them and make a correction for the offset.

Once we deal with vital effects, we are left primarily with the $\delta^{18}O$ changes related to temperature and ice volume. Colder temperatures and increasing ice volume both cause the $\delta^{18}O$ of foraminifera to rise, and warming or less ice cause $\delta^{18}O$ to fall. The $\delta^{18}O$ of Foraminifera thus gives us a composite picture of climate. Benthic Foraminifera live in the bottom waters of the oceans. As explained earlier, these waters originate as high-latitude surface waters. Therefore, benthic Foraminifera record high-latitude surface ocean temperatures along with ice volume and bottom water temperatures. Benthic foram $\delta^{18}O$ is as close as we can come to a single measure of global climate and serves to represent this property well.

There is an interesting history concerning attempts to partition the glacial-interglacial $\delta^{18}O$ change between temperature and ice volume. In one of his pioneering papers on the deep-sea oxygen isotope stratigraphy, Emiliani (1966) examined the data then available on the $\delta^{18}O$ of polar ice, and he concluded that temperature change dominated $\delta^{18}O$ variability. Shackleton (1967) ran an excellent experiment showing that glacial-interglacial changes of benthic forams were similar to those of planktonic forams. Since temperatures of benthic forams cannot drop too much (lest they fall below the freezing point), he

(*Box 2 continued*)

attributed the foram change primarily to a change in $\delta^{18}O$ of seawater. He thus concluded that $\delta^{18}O$ was primarily a signal of ice volume. In another landmark study, Adkins et al. (2002) accessed the $\delta^{18}O$ of the glacial deep ocean by analyzing the $\delta^{18}O$ of pore waters in sediments that, at $\sim$30 m depth, "remember" the composition of the glacial ocean. Their work, showing that the $\delta^{18}O$ of seawater rose by $\sim$1‰ during the ice age, slightly shifted things back toward Emiliani's ocean, and showed that bottom waters during the last glacial maximum were close to the freezing point.

Once deposited in sediments on the sea floor, Foraminifera recrystallize. In this process, the original small biogenic crystals slowly dissolve, reprecipitating as large $CaCO_3$ grains. Recrystallization proceeds because larger crystals are thermodynamically favored. The newly precipitated crystals will have $\delta^{18}O$ values reflecting the temperature and isotopic composition of the interstitial waters of the sediments, which can be very different from the conditions of formation. The problem is particularly severe for tropical planktonic Foraminifera, which grow at warm temperatures but will recrystallize at the cold temperatures of the sea floor. Benthic Foraminifera are less affected by this problem, and tend to be more robust of construction. By the Paleogene (Paleocene and Eocene), $\delta^{18}O$ values of planktonic Foraminifera were seriously altered, while the $\delta^{18}O$ record of benthic foraminifera are frequently reliable back through the Cenozoic. Occasionally, older sediments have been encountered in which even planktonic Foraminifera are very well preserved.

The new thing in the field is clumped isotope paleothermometry. This method takes advantage of the fact that heavy isotopes of oxygen and carbon are not randomly distributed in $CaCO_3$. Rather, they are "clumped," meaning that there are more molecules with heavy isotopes of both carbon (^{13}C) and oxygen (^{18}O) than one would expect from random distribution. The clumping is temperature dependent; it disappears at high temperatures, and thus an excess of clumped molecules is a measure of temperature. The great advantage of clumped isotope thermometry is that it allows one to assess temperature without any independent knowledge of the isotopic composition of the water from which the $CaCO_3$ precipitated. Clumping was only recently documented, and its application to a broad range of problems has begun. There will be many important applications of this method. However, it is limited to certain groups of organisms and inorganic $CaCO_3$ deposits because others are out of equilibrium for this property.

REFERENCES

Adkins, J. F., K. McInyre, and D. P. Schrag (2002), The salinity, temperature, and delta O-18 of the glacial deep ocean, *Science*, *298*, 1769–1773.

Emiliani, C. (1966), Paleotemperature analysis of Carribean cores P6304-8 and P6304-9 and a generalized temperature curve for the past 425,000 years, *The Journal of Geology*, *74*, 109–126.

Shackleton, N. (1967), Oxygen isotope analyses and Pleistocene temperatures re-assessed, *Nature*, *215*, 15–17.

been attributed to changes in Earth's orbit around the sun. This radiative forcing, discussed at length later in the book, including box 4, results from the fact that the eccentricity of Earth's orbit, the precession of the equinoxes, and the tilt of the spin axis vary with periods of approximately 20, 40, 100, and 400 Kyr. In a given region, orbital changes redistribute insolation between the summer and winter half years. According to the eponymous theory of Milutin Milankovitch, ice sheets grow when summers are cool, and melt when summers are warm.

Evidence for low CO_2 during the Late Paleozoic comes from the Geocarb models of Berner (2006), studies of the $\delta^{13}C$ of $CaCO_3$ precipitates in paleosols (Cerling et al. 1997), multiproxy studies of Royer et al. (2001), and the studies of stomatal size and density discussed earlier. Montanez et al. (2007) and Birgenheier et al. (2010) presented carbon isotope evidence for CO_2 fluctuations during the Permian. They then compared the timing of inferred low CO_2 levels with the timing of Permian glaciation. The results offer some evidence that atmospheric CO_2 burdens were higher during long periods when the planet was substantially ice free, but the details are complicated.

Evidence that changes in Earth's orbit around the sun caused climate change comes from the similarity between the periods of glacial cycles and the periods of orbital variations. Heckel (2008) counts 26 major cyclothems in the United States midcontinent between 311 and 301 Ma. The period is thus about 380 Kyr, close to one of the periods of eccentricity cycles, about 400 Kyr. Similarly, Davydov et al. (2010) showed that, between 307–314 Ma, there was also one major cyclothem every ~400 Kyr in the Donets Basin. In general, major cycles include several minor cycles. Minor cycles may correspond to shorter orbital periods, but we do not yet know because these smaller cycles are difficult to recognize and interpret in terms of global glaciation.

REMAINING MYSTERIES

It seems likely that the Late Paleozoic ice ages will eventually have a great deal to tell us about our current climate. At present there are some hints about fundamental questions like the link between glaciation and greenhouse gases, or the periodicity of glaciation. It is not known if the Late Paleozoic glaciations resembled the Cenozoic glaciations, during which, for the past 34 Myr, there has been a permanent ice sheet on Antarctica while lower latitude glaciation came and went. If so, the amplitude of sea level lowering would have been greater than estimated from facies changes associated with the cyclothems, which do not reflect the sea level contribution of persistent ice sheets.

A last question here concerns climate in the tropics. Isotopic paleotemperature studies suggest tropical Late Paleozoic temperatures around 21°–35°C (Angiolini et al. 2009; Came et al. 2007). For reference, equatorial temperatures range from about 18 to 30° today; low values are associated with coastal upwelling. However, there is evidence for cold climates in the tropics and subtropics during the late Paleozoic ice ages. As noted above, Brezinski et al. (2008) report Late Devonian glaciation in Pennsylvania; they estimate a paleolatitude of 30–45°. Sweet and Soreghan (2008) argue that ancient landforms imply seasonally freezing temperatures at low latitudes during the Late Paleozoic. Soreghan et al. (2009) interpret deposits of the Late Paleozoic Cutler Formation, southwestern United States, as proglacial sedimentation in the tropics. They interpret units of this formation as containing diamictites, glacially cut rocks, and dropstones. Furthermore, there is evidence for a low-elevation provenance.

In summary, evidence suggests that during the Late Paleozoic ice ages, the tropics were as warm as or warmer than today, but there is also apparently contradictory evidence for glaciation at low latitudes. If these interpretations are sustained, it will raise profound new questions about the nature of the Late Paleozoic ice ages.

REFERENCES

Papers with asterisks are suggested for further reading.

Angiolini, L., F. Jadoul, M. J. Leng, M. H. Stephenson, J. Rushton, S. Chenery, and G. Crippa (2009), How cold were the Early Permian glacial tropics? Testing sea-surface temperature using the oxygen isotope composition of rigorously screened brachiopod shells, *Journal of the Geological Society, London*, *166*, 933–945.

Berner, R. (2006), Geocarbsulf: A combined model for Phanerozoic atmospheric O_2 and CO_2, *Geochimica et Cosmochimica Acta*, *70*, 5653–5664.

Birgenheier, L. P., T. D. Frank, C. R. Fielding, and M. C. Rygel (2010), Coupled carbon isotopic and sedimentological records from the Permian system of eastern Australia reveal the response of atmospheric carbon dioxide to glacial growth and decay during the late Palaeozoic Ice Age, *Palaeogeography, Palaeoclimatology, Palaeoecology*, *286*, 178–193.

Bishop, J. W., I. P. Montanez, and D. A. Osleger (2010), Dynamic Carboniferous climate change, Arrow Canyon, Nevada, *Geosphere*, *6*, 1–34.

Brezinski, D., C. Cecil, V. Skema, and R. Stamm (2008), Late Devonian glacial deposits from the eastern United States signal an end of the mid-Paleozoic warm period, *Palaeogeography, Palaeoclimatology, Palaeoecology*, *268*, 143–151.

*Came, R. E., J. M. Eiler, J. Veizer, K. Azmy, U. Brand, and C. R. Weidman (2007), Coupling of surface temperatures and atmospheric CO_2 concentrations during the Palaeozoic era, *Nature*, *449*, 198–202.

Caputo, M. V., J. H. Goncalves de Melo, M. Streel, and J. L. Isbell (2008), Late Devonian and Early Carboniferous glacial records of South America, *Geological Society of America Special Papers*, *441*, 161–173.

Cerling, T. E., J. M. Harris, B. J. MacFadden, M. G. Leakey, J. Quade, V. Eisenmann, and J. R. Ehleringer (1997), Global vegetation change through the Miocene/Pliocene boundary, *Nature*, *389*, 153–158.

Chakraborty, C., and S. K. Ghosh (2008), Pattern of sedimentation during the Late Paleozoic, Gondwanaland glaciation: An example from the Talchir Formation, Satpura Gondwana basin, central India, *Journal of Earth System Science*, *117*, 499–519.

Davydov, V. I., J. L. Crowley, M. D. Schmitz, and V. I. Poletaev (2010), High-precision U-Pb zircon age calibration of the global Carboniferous time scale and Milankovitch band cyclicity in the Donets Basin, eastern Ukraine, *Geochemistry, Geophysics, Geosystems G³, An Electronic Journal of the Earth Sciences*, *11*, 1–22. doi: 10.1029/2009GC002736.

Falcon-Lang, H. J., W. J. Nelson, S. Elrick, C. V. Looy, P. R. Ames, and W. A. DiMichele (2009), Incised channel fills containing conifers indicate that seasonally dry vegetation dominated Pennsylvanian tropical lowlands, *Geology*, *37*, 923–926.

Fielding, C. R., T. D. Frank, and J. L. Isbell (2008), The late Paleozoic ice age—A review of current understanding and synthesis of global climate patterns, *Geological Society of America Special Papers*, *441*, 343–354.

Fielding, C. R., T. D. Frank, L. P. Birgenheier, M. C. Rygel, A. T. Jones, and J. Roberts (2008), Stratigraphic imprint of the Late Paleozoic Ice Age in eastern Australia: A record of alternating glacial and nonglacial climate regime, *Journal of the Geological Society, London*, *165*, 129–140.

*Frank, T. D., L. P. Birgenheier, I. P. Montanez, C. R. Fielding, and M. C. Rygel (2008), Late Paleozoic climate dynamics revealed by comparison of ice-proximal stratigraphic and ice-distal isotopic records, *Geological Society of America Special Papers*, *441*, 331–342.

Hampson, G., H. Stollhofen, and S. Flint (1999), A sequence stratigraphic model for the Lower Coal Measures (Upper Carboniferous) of the Ruhr district, north-west Germany, *Sedimentology*, *46*, 1199–1231.

Heckel, P. H. (1986), Sea-level curve for Pennsylvanian eustatic marine transgressive-regressive depositional cycles along midcontinent outcrop belt, North America, *Geology*, *14*, 330–334.

*Heckel, P. H. (2008), Pennsylvanian cyclothems in Midcontinent North America as far-field effects of waxing and waning of Gondwana ice sheets, *Geological Society of America Special Papers*, *441*, 275–289.

....

Isaacson, P., E. Diaz-Martinez, G. Grader, J. Kalvoda, O. Babek, and F. Devuyst (2008), Late Devonian–earliest Mississippian glaciation in Gondwanaland and its biogeographic consequence, *Palaeogeography, Palaeoclimatology, Palaeoecology*, *268*, 126–142.

Martin, J. R., J. Redfern, and J. F. Aitken (2008), A regional overview of the late Paleozoic glaciation in Oman, *Geological Society of America Special Papers*, *441*, 175–186.

Montanez, I. P., N. J. Tabor, D. Niemeier, W. A. DiMichele, T. D. Frank, C. R. Fielding, J. L. Isbell, et al. (2007), CO_2-forced climate and vegetation instability during Late Paleozoic deglaciation, *Science*, *315*, 87–91.

Mory, A. J., J. Redfern, and J. R. Martin (2008), A review of Permian-Carboniferous glacial deposits in Western Australia, *Geological Society of America Special Papers*, *441*, 29–40.

Royer, D., R. Berner, and D. Beerling (2001), Phanerozoic atmospheric CO_2 change: evaluating geochemical and paleobiological approaches, *Earth-Science Reviews*, *54*, 349–392.

Rygel, M. C., C. R. Fielding, T. D. Frank, and L. P. Birgenheier (2008), The magnitude of Late Paleozoic glacioeustatic fluctuations: A synthesis, *Journal of Sedimentary Research*, *78*, 500–511.

Soreghan, G. S., M. J. Soreghan, D. E. Sweet, and K. D. Moore (2009), Hot fan or cold outwash? Hypothesized proglacial deposition in the upper Paleozoic Cutler Formation, western tropical Pangea, *Journal of Sedimentary Research*, *79*, 495–522.

Sweet, D. E., and G. S. Soreghan (2008), Polygonal cracking in coarse clastics records cold temperatures in the equatorial Fountain Formation (Pennsylvanian-Permian, Colorado),

Palaeogeography, Palaeoclimatology, Palaeoecology, 268, 193–204.

Wilkinson, B. H., G. K. Merrill, and S. J. Kivett (2003), Stratal order in Pennsylvanian cyclothems, *Geological Society of America Bulletin, 115,* 1068–1087.

6 EQUABLE CLIMATES OF THE MESOZOIC AND PALEOGENE

....

EARTH WAS SUBSTANTIALLY UNGLACIATED DURING most or all of the interval beginning in the Upper Permian (260 Ma), through the Triassic, Jurassic, and Cretaceous periods of the Mesozoic Era, and the Paleocene and Eocene Periods of the Cenzoic Era, until the Eocene-Oligocene Boundary (34 Ma). Differences from modern climates were not limited to the absence of ice; the tropics were somewhat warmer than today, and the high latitudes were far warmer. While there were undoubtedly globally significant climate variations during the long interval of equable climates, the focus here is on the main characteristic of this time: Earth was warmer, especially at the high latitudes, in winter as well as summer. These very warm conditions are referred to as "equable climates."

Evidence for warm climates on land comes primarily from comparing the nature and temperature ranges of plant and animal fossils with modern climates in the study regions. Evidence for warm seawater temperatures comes from studies of the concentrations of organic chemicals, originating in ancient phytoplankton, whose relative concentrations depend on growth temperature. Ocean evidence also comes from the

....

isotopic composition of oxygen in fossil shells composed of $CaCO_3$ (as explained in box 2), and this chapter starts with these data. We aggregate evidence from different time periods, but the focus is largely on the Eocene, a very warm period whose younger age leads to a particularly rich climate record.

The striking difference between equable and modern climates has provoked modelers to examine the dynamics. It is possible to warm the planet simply by dialing up the atmospheric concentration of CO_2 and possibly CH_4. There is, indeed, evidence for higher CO_2 concentrations in the Eocene, Paleocene, and Cretaceous. However, changing greenhouse gases may not explain the dramatic warming of the high latitudes with respect to the tropics. Over the past decade, climate modelers have proposed a number of ways to explain this curious feature.

In this chapter, we examine evidence from marine and then terrestrial environments for equable climates during the Mesozoic Era, the Paleocene epoch, and the Eocene epoch, together spanning the interval from 245 to 34 Ma. We then discuss hypotheses that have been put forth to explain the dramatic warming with respect to the present day. We end with a summary of the limited evidence suggesting glaciation during this interval.

OCEAN TEMPERATURES DURING THE TIME OF EQUABLE CLIMATES

Our most important record of ocean temperatures for the last 100 Ma or so comes from data on the relative abundance of stable oxygen isotopes in the microscopic $CaCO_3$ skeleta of Foraminifera, a phylum of amoeboid protists. The causes and measurement of stable isotope abundance variations are discussed in box 1, and the significance to climate of measurements of the $\delta^{18}O$ of $CaCO_3$ is outlined in box 2. In a nutshell, increases in the $^{18}O/^{16}O$ ratio, or $\delta^{18}O$, of $CaCO_3$ signal some combination of colder temperatures and increased ice volume. Box 3 outlines other proxies that are widely used to study climates of the Cenozoic and earlier times.

The Cenozoic $\delta^{18}O$ record of benthic Foraminifera (living on the seafloor) is shown in figure 6.1 (Zachos et al. 2001). This iconic record integrates a large amount of data reflecting the main climate changes of the past 65 Myr in a parsimonious way. Changes in $\delta^{18}O$ are due to a combination of variations in temperature and changes in the size of the continental ice sheets. Changes in the volume of the ice sheets are linked to changes in sea level as glacial growth removes seawater or the decay of ice sheets returns water to the oceans. A $\delta^{18}O$ increase of 1‰ records a deep ocean cooling of about 4°C, an increase in continental ice volume associated with a sea level drop of 100 m, or some combination of the two. For reference, a 100 m fall in sea level corresponds to an average increase in the thickness of continental ice of about 200 m when spread over the entire continental area. Of course in practice, this ice would form a thicker sheet covering polar and temperate continental areas.

The data show that the warmest sustained period of the Cenozoic was around 52 Ma, at least insofar as high latitudes are concerned. The $\delta^{18}O$ of benthic forams

Box 3

Proxies for Characterizing Cenozoic Climates

Almost all information about past climates is based on proxies; these are properties that are not themselves climate properties, but nevertheless reflect past climates. Arguably the most important climate proxy is the stable oxygen isotope composition of $CaCO_3$ fossils (see box 2). However there are many other important proxies, and some are summarized here.

Many proxies are based on the nature of fossils. The temperature information in these data is obvious. Organisms have optimal temperature ranges, and the composition of fossil assemblages thus is an indication of ambient temperatures. In the oceans, assemblages of fossil Foraminifera and Radiolaria (both single-celled heterotrophic organisms) have been used extensively as temperature indicators. On land, plant assemblages have been the most important temperature indicators. Over the past 15–20 Ka, our most extensive information comes from pollen assemblages in ancient bogs or lake sediments. Records of tree-ring width go back over 10 Kyr at this time and give detailed records of regional climates. Turning to the oceans, micropaleontologists have developed statistical methods for quantitatively reconstructing temperatures from the abundances of microscopic fossil assemblages in deep-sea sediments.

Various chemical properties also serve as important proxies. Three have acquired particular significance for the reconstruction of ocean temperatures. The first is the Mg/Ca ratio of Foraminifera; this ratio increases as temperature rises. Second is the UK_{37} index. This property is derived from the relative abundance of alkenones, long-chained ketones with, in this case, 37 carbon atoms. Most carbon atoms in the chains are connected by a single pair of shared electrons. However, some carbon atoms are connected by two pairs of shared electrons.

The arrangement of these double bonds depends on temperature, and can be used to reconstruct past temperatures. Alkenones are produced by certain prymnesiophytes, a division of algae. The UK_{37} index reflects the temperature of the water near the sea surface, where these organisms live. A third chemical proxy is the TEX_{86} index. This property depends on the relative abundance of similar but distinct lipids making up the membranes of archaeal phytoplankton known as Crenarchaeota.

In addition to fossils, the nature of soils gives important information about past climates on land. Soils are formed by chemical decomposition, or "weathering" of the country rock. If climates are warm and wet, weathering will be more intense than if climates are cold and dry. Highly weathered soils therefore indicate warmer, wetter conditions. At the opposite end is loess, soil that forms from the settling of windblown dust. Loess signifies dry climates in the source area of the dust, and unweathered loess signifies relatively dry and cold conditions at the area of accumulation.

Over the past decade, speleothems (stalactites, stalagmites, and flowstones) from caves have become an increasingly critical source of information about continental climates. When the growth of a speleothem is intermittent, one infers that precipitation was greater during periods in which the specimen accreted than when there was no growth. In speleothems that have grown continuously, shifts in the $\delta^{18}O$ of the $CaCO_3$ signify shifts in the source and/or amount of precipitation. There are actually two reasons for changes in the $\delta^{18}O$ of a speleothem: the isotopic composition of precipitation could have changed, or temperature could have changed. In most speleothems studied to date, big isotopic changes have been observed. These must come from changes in the $\delta^{18}O$ of precipitation, because the required temperature changes would be improbably large. $\delta^{18}O$ of precipitation depends on the difference between evaporation temperature

(*Box 3 continued*)

and condensation temperature; it also depends on the seasonal and spatial distribution of rainfall. Speleothems have an extremely important ancillary benefit as well. Because they are massive, they are not affected by diagenesis. This attribute means that they can be dated to an accuracy limited only by our ability to make high-quality radioactive dating measurements. Speleothems are the most accurately dated Pleistocene climate deposit back to about 300 Ka, giving us a detailed account of regional hydrologic changes. Furthermore, since some of these changes are correlated to other events, we can transfer speleothem ages and thereby accurately date other climate records.

A series of deposits associated with glaciation gives essential climate information. Glaciers leave marks of their maximum extent in moraines (ridges of dirt and rock formed where glaciers melt), as well as in stream and lake deposits associated with melting. These deposits allow us to document the maximum extent of the great continental glaciers and the history of their retreat at the end of the last ice age. Moraines of mountain glaciers record regional climate changes leading to advances or retreats of the snowline. Glaciers calving into the oceans form icebergs. As icebergs melt, they release debris, which then accumulates in deep-sea sediments.

The proxies outlined above are among the most widely used in climate reconstructions, but the complete list is far longer. Developing climate proxies has been a wonderfully creative process, and a tremendous variety of properties has been exploited. Just two more examples give an idea of the range of properties employed. Beetle taxonomy is an important temperature proxy in some areas. Another is the concentrations in groundwater of noble gases, whose solubilities are temperature dependent; these gases have allowed the reconstruction of low-latitude continental paleotemperatures during the last ice age.

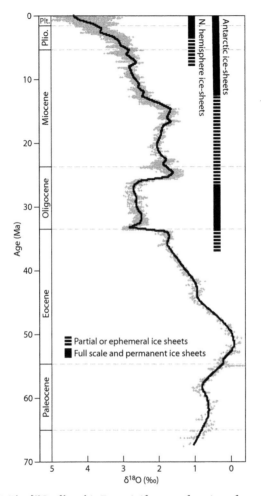

Fig. 6.1. The $\delta^{18}O$ of benthic Foraminifera, as a function of age during the Cenozoic (Zachos et al. 2001). Points to the right indicate warmer climates and smaller continental ice sheets.

today is about 3.0‰ heavier than at 52 Ma. Of this difference, ice present on Antarctica and Greenland accounts for about 0.6‰, or about 60 m of sea level. The remaining 2.4‰ increase in $\delta^{18}O$ of benthic forams signals that the deep ocean was about 10°C warmer in the Eocene than today. Cretaceous benthic forams, like those from the Eocene, also provide evidence for warm temperatures in the high latitudes (Moriya et al. 2007).

Deep water cooling is linked to a decrease in temperature of the high-latitude oceans. For obvious reasons, the deep ocean basins are filled with the densest water. As discussed in chapter 1, density depends on temperature and salinity. Surface waters at high latitudes are not the saltiest, but they are still the coldest and densest. Therefore, the high latitudes win the density war, and host the formation of dense waters that sink to fill the abyss. Temperatures of benthic Foraminifera in the deep sea are therefore similar to temperatures of surface waters at high latitudes. As described above, deep waters and, inferentially, high-latitude surface waters both cooled by about 10°C from the Eocene to the present. The temperature decrease of seawater is limited because, when the temperature falls to −2°, sea ice forms and there is no further cooling. Air temperature, not buffered by the formation of sea ice, cooled much more. This dramatic polar cooling was one of the most important climate changes of the Cenozoic, and is mirrored in a land cooling in the high latitudes, as we'll soon see.

A major effort has been invested in generating curves of Cenozoic surface water temperatures versus time in various ocean regions. The early results for low-latitude samples were completely unexpected: the $\delta^{18}O$ of planktonic Foraminifera actually indicated that during the Paleogene (34–65 Ma), tropical surface ocean temperatures were cooler than at present, rather than warmer (e.g., Bralower et al. 1995). It has since come to be understood that this result was due to diagenesis, the slow recrystallization of $CaCO_3$ through time: as the $CaCO_3$ crystals composing planktonic foram skeletons dissolved and reprecipitated on the seafloor in the millions of years since they grew, they acquired the imprint of cold bottom water temperatures in their oxygen isotope composition. Planktonic Foraminifera seem to be much more susceptible to this problem than the more robust benthics.

Recrystallization of planktonic Foraminifera was first demonstrated by Pearson et al. (2001) in a beautiful natural experiment. They analyzed Eocene Foraminifera "sealed" in impermeable clay layers from strata in Tanzania. The researchers showed that these forams had isotopic temperatures much warmer than those of the same foram species sampled from coarse-grained carbonates in permeable deep-sea sediments. Their conclusion was that the "sealed" forams retained their initial composition while Eocene forams in most deep-sea sediments had slowly recrystallized. Using similarly sealed samples, Moriya et al. (2007); Norris et al. (2002); and Wilson et al. (2002) documented temperatures of around 30°C in the tropical North Atlantic during the Cretaceous, between approximately 90–100 Ma. Throughout the Eocene, temperatures recorded from the Tanzanian sediments were ~32°C (Pearson et al. 2007). Planktonic Foraminifera from New Zealand give temperatures for the Early and Middle Eocene of up to 30°, despite the high paleolatitude—about 55° S (Hollis et al. 2009)! These warm temperatures in New Zealand were verified by the TEX$_{86}$ index, which registers temperature from the temperature-sensitive ratio of several slightly different organic compounds.

Additional evidence for warm, high-latitude oceans comes from remarkable fossils found on both Arctic islands and the Antarctic Peninsula. Temperatures of these small landmasses must reflect temperatures of the nearby oceans. Mean annual temperatures derived for the Antarctic Peninsula between about 90–40 Ma range from 10 to 20°C; these numbers come from the presence and wood structure of fossil trees (Poole et al. 2005). On Spitsbergen and northern Ellesmere Island, there were rich forests in the Cretaceous that included ginkgos, ferns, angiosperms, and conifers (Falcon-Lang et al. 2004; Harland et al. 2007). Even more striking, during the Eocene, lizards (Varanidae), turtles, and alligators lived on Ellesmere Island, which was located immediately west of northern Greenland (Estes and Hutchison 1980)! Modern alligators require a mean annual temperature above 15°C, and a monthly mean cold temperature above 5°C. Modern varanids require even warmer conditions. The Eocene paleolatitude of Ellesmere Island was about the same as today, 78° N. The presence of these animals is therefore quite stunning.

LAND TEMPERATURES DURING THE CRETACEOUS AND PALEOGENE

As one might expect from ocean temperatures, the land flora and fauna indicate much warmer temperatures than today in midlatitude and temperate regions. We'll review some of the evidence, starting in the Arctic. On the North Slope of Alaska, where today we find tundra, mid-Cretaceous fossils indicate that forests were present containing ferns, ginkos, agiosperms, and conifers, much as on Ellesmere Island (Spicer and Parrish 1986; Spicer and Herman 2001). Dinosaurs lived in the Arctic during the Cretaceous. This in itself is not surprising, because dinosaurs (birds) occupy these areas today. However, the Cretaceous inhabitants included groups that were probably cold-blooded, such as hadrosaurs in the Yukon, and stegosaurs and sauropods in Siberia (Rich et al. 2002; Seebacher 2003); these animals would have required much higher temperatures than exist in the regions today. In the midnorthern latitudes, the most dramatic evidence for warmer climates comes from crocodiles. During the Eocene, these were present in the region of the US mountain states up to a paleolatitude of 55° N. Alligators today live up to about 35° N.

Fossil plants and animals from the tropics generally don't provide clear information about equable climates: How could we tell, when comparing fossil with modern organisms, if tropical temperatures were warmer in the past? Recently, a Paleocene snake was discovered in northeastern Colombia that was estimated to be 13 m in length, and weigh about 1100 kg (Head et al. 2009). The authors estimated the ambient temperature from the relationship between skeletal properties of modern snakes and temperature, and concluded that this tropical region was about 5°C warmer in the Paleocene than today. This approach turned out to be controversial and the conclusions are under debate.

Heading south, plant fossils indicate that Eocene temperatures at a site in Patagonia (Laguna del Hunco, Chubut Province, Argentina) were about 4°C warmer than today (Wilf et al. 2003). Eocene pollen from Pridz Bay, on the coast of Antarctica in the Indian Ocean sector, suggests a climate supporting taiga, including dwarf trees, herbs, and shrubs (Truswell and Macphail 2009). Jurassic and Cretaceous dinosaur fossils have also been found in Antarctica, including prosauropods, which were likely cold-blooded (Rich et al. 2002; Seebacher 2003).

The meridional gradient in Middle Eocene mean annual temperatures is summarized in figure 6.2 (Greenwood and Wing 1995). The gray band in each panel shows the range of modern average annual temperature versus latitude (top) or cold month mean temperature (bottom). Recent foram oxygen isotope data indicate tropical sea surface temperatures around 35°C (Moriya et al. 2007; Pearson et al. 2001; Norris et al. 2002; Pearson et al. 2007; Wilson et al. 2002). Overall, results suggest that tropical temperatures were elevated by a few degrees, midlatitude temperatures by somewhat more, and temperatures in subpolar and polar regions by 10°C

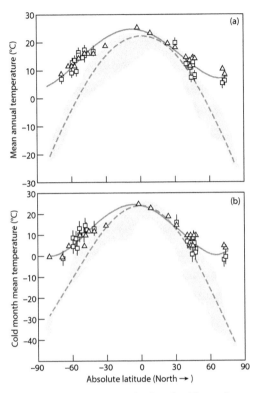

Fig. 6.2. Mean annual temperature (*top*) and cold month mean temperature (bottom) versus latitude during the Eocene (from Greenwood and Wing 1995)). The gray band in each figure represents the modern condition. Open squares and closed triangles represent land temperatures, estimated in different ways from fossil vegetation.

or much more. The lower plot of this figure shows that cold month mean temperatures were quite equable, remaining above the freezing point even around 80° latitude.

DYNAMICS OF EQUABLE CLIMATES

The challenge is to explain two salient facts of Mesozoic and Paleogene climates: Earth's average temperature was significantly higher, and the high latitudes warmed considerably more than the tropics. One can account for the global warming by invoking some combination of higher greenhouse gas concentrations and lower albedo. A number of ideas has been advanced to explain the high-latitude warming. One is that more heat was transported from the tropics toward the poles. A second is that increased Paleogene vegetation lowered the albedo of high-latitude continents. A third is that wintertime clouds at high latitudes would have exerted a strong greenhouse warming. Below, we discuss these ideas in turn.

Data in figure 6.2 indicate that, during the Eocene, Earth was warmer by ~5°C in the tropics and 30°C or more at certain high-latitude locations. In calculating global temperature, we need to weight the low latitudes more, since half of Earth's surface is within 30° of the equator. Thus, a reasonable estimate for Eocene warming compared with modern conditions is ~7°C based on the reconstruction of Greenwood and Wing (1995), and ~14°C based on Huber and Caballero 2011). In chapter 4, we examined proxy evidence indicating that Eocene atmospheric CO_2 levels were three to six times higher than modern, although greater increases cannot be ruled out (Huber and Caballero 2011). Modeling studies aimed at assessing the consequences of global change suggest that the planet warms by about 2.5°C (with a large uncertainty) given a doubling in CO_2. This number suggests that CO_2 would lead to an Eocene warming of about 6°C (again with a large uncertainty). If CO_2 were 3000 ppm (approximately 11 × preindustrial), the warming would be about 9°C.

There was an additional warming, because the albedo was lower; Earth was ice free, and deserts were largely replaced by forests. Hence, explaining the increased global average temperature during the Eocene seems fairly straightforward. Studies using models with Eocene boundary conditions reinforce this conclusion (Shellito et al. 2003; Sloan et al. 1995), although the models may require slightly more than a quadrupling of CO_2 to raise global temperature by at least 7°C.

However, many modeling studies have failed to simulate the exceptional warming observed for the high latitudes (Barron 1987; Shellito et al. 2003; Sloan and Barron 1992). This result has prompted a search for climate processes that may have been important during the Eocene, but are not properly captured in some relevant models. Initially, attention was on meridional heat transport; the simplest way to warm the high latitudes is to transport heat, through the ocean or atmosphere, from the tropics. This process is active and important today. If it was more intense in the past, it would have lead to additional high-latitude warming. However, a number of modeling studies showed that the mechanism does not appear to be viable (Barron 1987; Huber and Sloan 2001; Huber et al. 2004; Shellito et al. 2003; Sloan and Barron 1992). The basic problem is that the rate of meridional heat transport depends on the temperature gradient. As the high latitudes warm, and temperatures become more similar at low and high latitudes, it becomes more difficult to transfer heat from the tropics to the poles. In climate simulations, models can match high-latitude warming if CO_2 is raised to high enough values. However, some of these models face two problems: the tropics overheat, and the required CO_2 levels are greater than some proxies suggest.

Dynamicists have isolated other processes and examined whether, along with elevated concentrations of CO_2, they could contribute to the exceptional high-latitude warming. Najjar et al. (2002) examined changes in ocean circulation resulting from the different positions of the continents during the Eocene. They found that the Northern Hemisphere warmed because of changes in the shallower, wind-driven circulation, and the Southern Hemisphere warmed because of changes in the formation of deep ocean waters at high latitudes. Brady et al. (1998) found, in their simulation of Cretaceous ocean circulation, that warm, salty surface waters forming in the Southern Hemisphere subtropics flowed poleward to sink between 60° S and the Antarctic coast. While the waters cool and perhaps become slightly fresher on their way south, the simulations showed that they stayed relatively warm, with Antarctic sea surface temperatures around 10°C.

Korty et al. (2008) revisited the question of meridional heat transport, focusing on the role of tropical cyclones (large storms, including hurricanes). They argued that cyclones would be more intense in the warmer climates of a high-CO_2 world. An immediate effect of a tropical storm is to vertically mix the local ocean. Since subsurface water is cooler and denser, subsurface waters warm while surface waters cool. In the aftermath of storms, these cooler surface waters are warmed by the sun, and the average temperature of waters in the upper few hundred meters of the oceans increase. When these waters are transported to the high latitudes, they carry correspondingly more heat, increasing the meridional heat flux. The overall result is to warm the high-latitude ocean by 1–2°C, and to cool the tropics by a smaller amount.

Upchurch et al. (1998) investigated another effect, the lower albedo due to the land biosphere. They noted that Earth was vegetated much more heavily during Cretaceous times than at present (fig. 6.3), and diagnosed the effect of the change in albedo. Fully vegetated land absorbs light much more efficiently than desert or sparsely vegetated surfaces, and leads to warming. They calculated that the Eocene vegetation would induce warming of 4–8C° in Arctic continental regions relative to ground free of vegetation. The actual Eocene warming would have been smaller because modern land is not free of plants.

Two possibilities have been suggested for warming high latitudes by increasing wintertime cloud cover. Clouds have opposing effects on surface temperature. First, they reflect sun's light and cool the surface. Second, they have greenhouse properties, trapping Earth's outgoing long-wave radiation and reradiating it toward the surface. This latter effect explains why summer nights

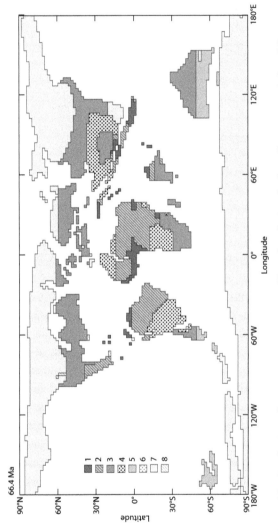

Fig. 6.3. The distribution of vegetation types over Earth's surface during the Late Cretaceous as inferred from a model guided by data. (1) Tropical rainforest; (2) tropical semideciduous forest; (3) subtropical broad-leaved evergreen forest and woodland; (4) desert and semidesert; (5) temperate evergreen broad-leaved and coniferous forest; (6) tropical savannah (not used here); (7) polar deciduous forest; (8) bare soil. A large area of the continents currently covered by grasslands and deserts was forested during the Late Cretaceous (from Upchurch Jr. et al. 1998).

can be quite cool when the sky is clear. At most places and times, reflection wins, and clouds lead to cooler surface temperatures. At high latitudes during winter, sunlight is dim or absent, and albedo does not have an important impact on climate. However, clouds still exert a greenhouse effect. As a consequence, more clouds lead to warmer winters in polar regions.

One route to increased wintertime cloudiness lies in the troposphere. If the Arctic Ocean were warmer (due, for example, to elevated CO_2), it would be ice free during the winter. Water would evaporate, and air above the sea surface would warm by contact with the warmer sea. This warm air would be buoyant; it would rise, cool, and vapor would condense as clouds (Abbot and Tziperman 2008; Abbot et al. 2009). According to modeling studies of Abbot et al. (2009), greenhouse forcing over the Arctic would increase by about 40 W m^{-2} due to winter clouds. Global average incoming solar radiation, after accounting for Earth's albedo, is about 235 W m^{-2}, and an increase of 40 W m^{-2} during winter at high latitudes is significant. The increased greenhouse forcing over nearby continents is smaller, of order 20 W m^{-2}, but still important. Hence, increased convection due to an ice-free Arctic could result in dramatic warming of the oceans as well as the high-latitude continents.

A second route to increased cloudiness lies in the stratosphere. Normally the stratosphere is cloud free because it is sourced by very cold (hence dry) air at the tropical tropopause, and because temperature within the stratosphere rises with altitude. Today, the stratosphere becomes cold enough during polar night that water condenses to form "polar stratospheric clouds," but these are thin and have weak greenhouse effects. If, however, the concentration of CH_4 were higher, the water concentration of the stratosphere would rise as a result of methane oxidation ($CH_4 + 2\,O_2 \rightarrow CO_2 + 2\,H_2O$). Sloan and Pollard (1998) invoke this process, speculating that warmer, wetter climates would have led to more wetlands and greater CH_4 production. They show that if the resulting clouds were sufficiently thick during the Eocene, they would exert a strong radiative forcing in high northern latitudes.

Alternatively, Kirk-Davidoff et al. (2002) suggest that dynamic changes would lead to more cloudiness in the winter polar stratosphere. They argue that higher Paleogene CO_2 would lead to warmer climates that would, in turn, slow down the circulation of the stratosphere. Air entering the stratosphere in the tropics would be warmer and would thus have a higher vapor content. On the other hand, the polar stratosphere would be colder, and thicker polar stratospheric clouds would condense. A simulation showed that this mechanism could deliver a radiative forcing of about 15 W m^{-2}, which again would be quite significant.

Recently, Huber and Caballero (2011) showed that CO_2 concentrations of 4500 ppm could account for much of the proxy temperature data describing Eocene climates. Their model simulated some of the specific feedbacks described above. Understanding whether CO_2 concentrations ever reached these levels, and whether temperatures in the tropics reached simulated values of 35–40° or more, is a task for future research.

WAS EARTH EVER GLACIATED DURING CRETACEOUS AND PALEOGENE TIMES OF EQUABLE CLIMATES?

A number of studies have argued for significant global glaciation during the Cretaceous and Paleogene. If correct, this work poses a challenge to our understanding of periods of equable climates.

Gale et al. (2002), Kominz et al. (2008), and Miller et al. (2009) examined long records of coastal sedimentation on sinking continental margins off New Jersey, Delaware, and southern India. They interpreted changes in sediment lithology as indicating changes in sea level which were, in turn, attributed to the growth and decay of major ice sheets. Essentially, these authors believe that they identified Cretaceous cyclothems.

Bornemann et al. (2008) and Tripati et al. (2005) both diagnose glaciation from the $\delta^{18}O$ of foraminiferal $CaCO_3$. Bornemann et al. identify an episode during the Turonian (~90 Ma), where planktonic and benthic foram data both indicate a combined cooling/increase in ice volume on the Demarara Rise. Using the TEX_{86} proxy to estimate the temperature change, they were able to show that the $\delta^{18}O$ of surface water had changed during this event, and concluded that the same was true of the deep water. This ocean $\delta^{18}O$ change would imply a global sea level change of about 50 m. Tripati et al. (2005) present similar evidence for an event in the Middle Eocene, at about 41 Ma, and estimate a sea level change of about 150 m. For reference, the glacioeustatic sea level lowering is 60 m today; during the Pleistocene, it was up to 190 m relative to an ice-free planet. Other studies have indicated the absence of variability in foram isotopes during the key intervals (Ando et al. 2010; Moriya et al. 2007).

Judging by today's ice sheets, Greenland and Antarctica could have accommodated about 60 m of ice, and the evidence might be lost by glacial erosion or hidden below the current glaciers. Sea level lowering of 150 m requires glaciation of currently unglaciated regions. This ice volume is much greater than Earth experienced during much of the late Paleozoic. There is extensive direct evidence for late Paleozoic glaciation, and we would expect to find direct evidence for 150 m of Eocene glaciation as well (diamictites, moraines, scored and polished pavement, etc.). Alley and Frakes (2003) identified a Lower Cretaceous diamictite (a cemented, poorly sorted rock deposit) from the Flinders Range, Australia, as a deposit of likely glacial origin. However, the deposit can probably be attributed to local glaciation at its paleolatitude of 66° S. Direct evidence of large continental glaciers during times of equable climates appears to be lacking. In our view the intriguing question of globally significant Cretaceous-Paleogene glaciation remains to be settled.

REFERENCES

Papers with asterisks are suggested for further reading.

Abbot, D. S., and E. Tziperman (2008), A high-latitude convective cloud feedback and equable climates, *Quarterly Journal of the Royal Meteorological Society*, *134*, 165–185.

Abbot, D. S., C. C. Walker, and E. Tziperman (2009), Can a convective cloud feedback help to eliminate winter sea ice at high CO_2 concentrations? *Journal of Climate 22*, 5719–5731.

Alley, N. F., and L. A. Frakes (2003), First known Cretaceous glaciation: Livingston Tillite Member of the Cadna-owie formation, South Australia, *Australian Journal of Earth Sciences*, *50*, 139–144.

Ando, A., B. T. Huber, K. G. MacLeod, T. Ohta, and B.-K. Khim (2010), Blake Nose stable isotopic evidence against the mid-Cenomanian glaciation hypothesis, *Geology*, *37*, 451–454.

Barron, E. J. (1987), Eocene Equator-To-Pole surface ocean temperatures: A significant climate problem? *Paleoceanography*, *2*, 729–739.

Bornemann, A., R. D. Norris, O. Friedrich, B. Beckmann, S. Schouten, J.S.S. Damste, J. Vogel, et al. (2008), Isotopic evidence for glaciation during the Cretaceous supergreenhouse, *Science*, *319*, 189–192.

Brady, E. C., R. M. DeConto, and S. L. Thompson (1998), Deep water formation and poleward ocean heat transport in the warm climate extreme of the Creaceous (80 Ma), *Geophysical Research Letters*, *25*, 4205–4208.

Bralower, T. J., J. C. Zachos, E. Thomas, M. Parrow, C. K. Paull, D. C. Kelly, I. P. Silva, et al. (1995), Late Paleocene to Eocene paleoceanography of the equatorial Pacific Ocean: Stable isotopes recorded at Ocean Drilling Program Site 865, Allison Guyot, *Paleoceanography*, *10*, 841–865.

Estes, R., and J. H. Hutchison (1980), Eocene lower vertebrates from Ellesmere Island, Canadian Arctic Archipelago, *Palaeogeography, Palaeoclimatology, Palaeoecology*, *30*, 325–347.

Falcon-Lang, H. J., R. A. MacRae, and A. Z. Csank (2004), Palaeoecology of Late Cretaceous polar vegetation preserved in the Hansen Point Volcanics, NW Ellesmere Island, Canada, *Palaeogeography, Palaeoclimatology, Palaeoecology*, *212*, 45–64.

Gale, A. S., J. Hardenbol, B. Hathway, W. J. Kennedy, J. R. Young, and V. Phansalkar (2002), Global correlation of Cenomanian (Upper Cretaceous) sequences: Evidence for Milankovitch control on sea level, *Geology*, *30*, 291–294.

*Greenwood, D. R., and S. L. Wing (1995), Eocene continental climates and latitudinal temperature gradients, *Geology*, *23*, 1044–1048.

Harland, M., J. E. Francis, S. J. Brentnall, and D. J. Beerling (2007), Cretaceous (Albian-Aptian) conifer wood from Northern Hemisphere high latitudes: Forest composition and palaeoclimate, *Review of Palaeobotany and Palynology*, *143*, 167–196.

Head, J. J., J. I. Bloch, A. K. Hastings, J. R. Bourque, E. A. Cadena, F. A. Herrera, P. D. Polly, and C. A. Jaramillo (2009), Giant boid snake from the Palaeocene neotropics reveals hotter past equatorial temperatures, *Nature*, *457*, 715–717.

Hollis, C. J., L. Handley, E. M. Crouch, H.E.G. Margans, J. A. Baker, J. Creech, K. S. Collins, et al. (2009), Tropical sea temperatures in the high-latitude South Pacific during the Eocene, *Geology*, *37*, 99–102.

Huber, M., and L. C. Sloan (2001), Heat transport, deep waters, and thermal gradients: Coupled simulation of an Eocene Greenhouse Climate, *Geophysical Research Letters*, *28*, 3481–3484.

Huber, M., and R. Caballero (2011), The early Eocene equable climate problem revisited, *Climate of the Past*, *7*, 603–633.

Huber, M., H. Brinkhuis, C. E. Stickley, K. Doos, A. Sluijs, J. Warnaar, S. A. Schellenberg, and G. L. Williams (2004), Eocene circulation of the Southern Ocean: Was Antarctica kept warm by subtropical waters? *Paleoceanography*, *19*, 1–12.

Kirk-Davidoff, D. B., D. P. Schrag, and J. G. Anderson (2002), On the feedback of stratospheric clouds on polar climate, *Geophysical Research Letters*, *29*, 1–4.

Kominz, M. A., J. V. Browning, K. G. Miller, P. J. Sugarman, S. Mizintseva, and C. R. Scotese (2008), Late Cretaceous to Miocene sea-level estimates from the New Jersey and Delaware coastal plain coreholes: An error analysis, *Basin Research*, *20*, 211–226.

Korty, R. L., K. A. Emanuel, and J. R. Scott (2008), Tropical cyclone-induced upper-ocean mixing and climate: Application to equable climates, *Journal of Climate*, *21*, 638–654.

Miller, K. G., P. J. Sugarman, J. V. Browning, M. A. Kominz, J. C. Hernandez, R. K. Olsson, J. D. Wright, et al. (2009), Late Cretaceous chronology of large, rapid sea-level changes: Glacioeustasy during the greenhouse world, *Geology*, *31*, 585–588.

Moriya, K., P. A. Wilson, O. Friedrich, J. Erbacher, and H. Kawahata (2007), Testing for ice sheets during the mid-Cretaceous greenhouse using glassy foraminiferal calcite from the mid-Cenomanian topics on Demerara Rise, *Geology*, *35*, 615–618.

Najjar, R. G., G. T. Nong, D. Seidov, and W. H. Peterson (2002), Modeling geographic impacts on early Eocene ocean temperature, *Geophysical Research Letters*, *29*, 1–4.

*Norris, R. D., K. L. Bice, E. A. Magno, and P. A. Wilson (2002), Jiggling the tropical thermostat in the Cretaceous hothouse, *Geology*, *30*, 299–302.

Pearson, P. N., B. E. van Dongen, C. J. Nicholas, R. D. Pancost, S. Schouten, J. M. Singano, and B. S. Wade (2007), Stable warm tropical climate through the Eocene Epoch, *Geology*, *35*, 211–214.

*Pearson, P. N., P. W. Ditchfield, J. Singano, K. G. Harcourt-Brown, C. J. Nicholas, R. K. Olsson, N. J. Shackleton, and M. A. Hall (2001), Warm tropical sea surface temperatures in the Late Cretaceous and Eocene epochs, *Nature* 413, 481–488. http://www.nature.com.

Poole, I., D. Cantrill, and T. Utescher (2005), A multi-proxy approach to determine Antarctic terrestrial palaeoclimate during the Late Cretaceous and Early Tertiary, *Palaeogeography, Palaeoclimatology, Palaeoecology*, *222*, 95–121.

*Rich, T. H., P. Vickers-Rich, and R. A. Gangloff (2002), Polar Dinosaurs, *Science*, *295*, 979–980.

Seebacher, F. (2003), Dinosaur body temperatures: The occurrence of endothermy and ectothermy, *Paleobiology*, *29*, 105–122.

Shellito, C. J., L. C. Sloan, and M. Huber (2003), Climate model sensitivity to atmospheric CO_2 levels in the Early-Middle Paleogene, *Palaeogeography, Palaeoclimatology, Palaeoecology*, *193*, 113–123.

Sloan, L. C., and E. J. Barron (1992), A comparison of Eocene climate model results to quantified paleoclimatic interpretations, *Palaeogeography, Palaeoclimatology, Palaeoecology*, *93*, 183–202.

Sloan, L. C., and D. Pollard (1998), Polar stratospheric clouds: A high latitude warming mechanism in an ancient greenhouse world, *Geophysical Research Letters*, *25*, 3517–3520.

Sloan, L. C., J.C.G. Walker, and T. C. Moore, Jr. (1995), Possible role of oceanic heat transport in early Eocene climate, *Paleoceanography*, *10*, 347–356.

Spicer, R. A., and J. T. Parrish (1986), Paleobotanical evidence for cool north polar climates in middle Cretaceous (Albian-Cenomanian) time, *Geology*, *14*, 703–706.

Spicer, R. A., and A. B. Herman (2001), The Albian-Cenomanian flora of the Kukpowruk River, western North Slope, Alaska: Stratigraphy, palaeofloristics, and plant communities, *Cretaceous Research*, *22*, 1–40.

Tripati, A., J. Backman, H. Elderfield, and P. Ferretti (2005), Eocene bipolar glaciation associated with global carbon cycle changes, *Nature*, *436*, 341–346.

Truswell, E. M., and M. K. Macphail (2009), Polar forests on the edge of extinction: What does the fossil spore and pollen evidence from East Antarctica say? *Australian Systematic Botany*, *22*, 57–106.

Upchurch, G. R., Jr., B. L. Otto-Bliesner, and C. Scotese (1998), Vegetation-atmosphere interactions and their role in global warming during the latest Cretaceous, *Philosophical Transactions of the Royal Society B-Biological Sciences*, *353*, 97–112.

Wilf, P., N. R. Cuneo, K. R. Johnson, J. F. Hicks, S. L. Wing, and J. D. Obradovich (2003), High plant diversity in Eocene South America: Evidence from Patagonia, *Science*, *300*, 122–125.

Wilson, P. A., R. D. Norris, and M. J. Cooper (2002), Testing the Cretaceous greenhouse hypothesis using glassy foraminiferal calcite from the core of the Turonian tropics on Demerara Rise, *Geology*, *30*, 607–610.

*Zachos, J. C., M. Pagani, L. C. Sloan, E. Thomas, and K. Billups (2001), Trends, rhythms, and aberrations in global climate 65 Ma to present, *Science*, *292*, 686–693.

···

THE PALEOCENE-EOCENE THERMAL MAXIMUM (PETM) was an event of about 200 Kyr duration, starting at the onset of the Eocene. During the event, a massive amount of CO_2 was rapidly released to the oceans and atmosphere. An alternative scenario for the kickoff is that CH_4 was released, and rapidly oxidized to CO_2. The consequences were predictable and dramatic. Global temperatures rose by about 6°C. The oceans became acidified, perhaps contributing to the extinction of many deep-dwelling calcareous organisms. Also, caustic seawater dissolved $CaCO_3$ shells that normally accumulate on the seafloor at intermediate ocean depths. On land, warming led to changes in the distribution of precipitation, an increase in weathering rates in some areas, and changes in the flora and fauna. While this general response is well documented, the origin(s) of the greenhouse releases that initiated this event remain a subject of speculation and analysis.

The magnitude of the CO_2 release is estimated to be between 2000 and 6000 Gt C (1 Gt C $= 10^9$ tons C $= 10^{14}$ gm C). The range is comparable to the anthropogenic CO_2 release if we burn all available coal, oil, and natural gas. In addition to global warming and ocean

···

acidification, the PETM response involved neutralization of CO_2 by $CaCO_3$ sediments, changes in the hydrological cycle, and the responses of land and ocean ecosystems to climate change, all of which are of interest in the context of anthropogenic global change. As a remarkable event in nature that is also something of an analog for the impending high-CO_2 world, the PETM has sparked great interest.

The PETM event was first identified by Kennett and Stott (1991). In a Southern Ocean core, they found large transient decreases in $\delta^{13}C$ and $\delta^{18}O$ in Foraminifera skeletons at the Paleocene-Eocene boundary. The $\delta^{18}O$ decrease signaled a warming of about 5°C, while the $\delta^{13}C$ decrease signified the addition of a large amount of biological CO_2 to the oceans. The isotopic carbon composition indicates a biological pedigree because organisms discriminate against ^{13}C, and in favor of ^{12}C, when making organic matter from CO_2. This discrimination is anywhere from 5 to 25‰ (but mostly toward the higher value), which is to say that the $\delta^{13}C$ of organic matter is ~25‰ less than that of CO_2 in air or dissolved inorganic carbon in seawater. In addition, methanogenic bacteria decompose organic matter into CO_2 and CH_4. There is a further isotope fractionation in this step, such that biogenic CH_4 has a $\delta^{13}C$ of about −60‰. So the change in oceanic $\delta^{13}C$ found by Kennett and Stott implies addition of biogenic carbon, and the required magnitude is about half as great if CH_4 rather than CO_2 is the ultimate source.

Since the initial work of Kennett and Stott, the foram $\delta^{13}C$ and $\delta^{18}O$ decreases have been identified in cores from the Atlantic, Pacific, Indian, Southern, and Arctic Oceans at the time of the Paleocene-Eocene boundary; see original data and summaries in Handley et al. (2008); Nunes and Norris (2006); Rohl et al. (2007); Weijers et al. (2007); Zachos et al. (2008); Zachos et al. (2007); and Zachos et al. (2006). The $\delta^{13}C$ decrease is also present on land, where (for example) it is recorded in ancient soils of Paleocene-Eocene age in the North American mountain west (Wing et al. 2005).

In this chapter, we start by discussing the carbon isotope data that record an extraordinary input of CO_2 or CH_4 into the ocean or atmosphere at the boundary between the Paleocene and Eocene epochs. We then discuss evidence for the magnitude of the CO_2 release to the oceans and atmosphere, the magnitude of the associated warming, and the imprint of the CO_2 release in the $CaCO_3$ abundance of deep-sea sediments. We end with a summary of hypotheses to account for the release.

THE PETM EVENT: TIMING, DURATION, AND MAGNITUDE OF THE TEMPERATURE CHANGE

As noted above, carbon isotope records, from both land and oceans, reflect the timing, intensity, and duration of the PETM. Figure 7.1 shows the $\delta^{13}C$ and $\delta^{18}O$ changes as observed in deep-sea foram records. *Nuttalides trumpeyi* is a benthic (bottom-dwelling) foram, and records conditions in the deep sea; the other two species are planktonic (surface dwelling), and record conditions at the sea surface. Figure 7.2 shows the $\delta^{13}C$

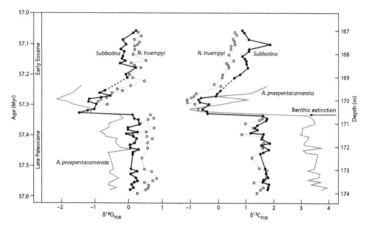

Fig. 7.1. The δ^{13}C and δ^{18}O of the planktonic foraminifera *Acarinina praepentacamerata* and *Subbotina*, and benthic foram *Nuttalides truempyi*, plotted against age and depth, in Ocean Drilling Program hole 690B (Kennett and Stott 1991). The large δ^{13}C decrease at 170.6 m depth observed in all three species marks the beginning of the Paleocene-Eocene Thermal Maximum. The original chronology (shown on this figure) has been revised and the δ^{13}C transition is now known to occur immediately after the Paleocene-Eocene boundary, in the earliest Eocene, at the depth marked Benthic extinction.

change as observed in sedimentary organic matter from paleosols of two Wyoming sites (from Nunes and Norris 2006; Wing et al. 2005). Both foram and paleosol records show the large decrease in δ^{13}C at the Paleocene-Eocene boundary. One noteworthy feature of the benthic foram records is the almost complete absence of any intermediate δ^{13}C values at the onset of the event. More such evidence comes from the work of Zachos et al. (2008), who analyzed single Foraminifera shells to avoid the problem

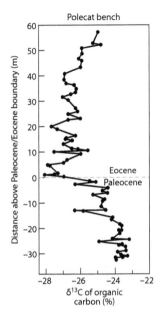

Fig. 7.2. $\delta^{13}C$ versus depth above and below the Paleocene-Eocene boundary for organic matter in paleosols averaged for two sites in Wyoming (Wing et al. 2005).

that mixing of specimens growing before and after the onset of the PETM can give intermediate isotope values. In samples from several deep-sea sites and a New Jersey coastal site, Zachos et al. (2008) found that the isotope composition of individual Foraminifera was bimodal; tests either had values corresponding to conditions before the shift or after the shift, but no intermediate values.

The abruptness of the shift must result from two factors. First, the onset of the PETM occurred in a very short time. Second, the addition of CO_2 to the oceans dissolved $CaCO_3$, making the transition in some places appear even more rapid than it really was. The period over which CO_2 was added, extending from the previous $\delta^{13}C$ baseline to the time of minimum $\delta^{13}C$, has been estimated to be about 10,000 years.

The duration of the entire PETM is thought to have been about 200 Kyr. This estimate is based on two lines of evidence. First, shorter climate cycles in the PETM were superimposed on the larger, rapid change and long recovery. These small cycles have been related to climate forcing from the periodic change in the position of the equinoxes in Earth's orbit around the sun. The equinoxes vary on a 20,000 year cycle, thereby providing a ticking clock for telling time (Rohl et al. 2007). Second, the abundance of "interplanetary dust particles" that are thought to rain out of the sky at a roughly constant rate also provides time information (Farley and Eltgroth 2003).

The temperature rise seems to have been about 5°C in the tropics and 8°C at high latitudes, with considerable uncertainty. Warming in the deep sea, inferred from the $\delta^{18}O$ of benthic forams, was typically about 5°C (Nunes and Norris 2006). The warming was likely similar at the high-latitude sea surface from whence the deep waters came. Continental warming was similar in midlatitudes of North America (Wing et al. 2005). It was also similar or somewhat larger in Arctic regions (Sluijs et al. 2006; Weijers et al. 2007).

The intensity of the event, as recorded both in foram C and O isotopes, was greatest at its inception, then tailed off to background levels with a half-time of about 100 Kyr. This number is similar to the residence time of outgassed CO_2 in the modern ocean-atmosphere system.

It is thought to reflect the rate at which excess CO_2, somehow injected to initiate the event, was consumed by weathering. The duration of the PETM was thus determined by the time required for weathering to take up the huge CO_2 pulse introduced at the start.

THE MAGNITUDE OF CO_2 INJECTION AND GLOBAL WARMING

A focus of PETM studies involves the attempt to rationalize this event in the context of our paradigm for understanding the modern carbon cycle and its link to climate. The idea is to assemble a falsifiable coherent narrative linking changes in the seawater CO_2 concentration, ocean acidity (as reflected by the preservation of $CaCO_3$ in sediments), $\delta^{13}C$ of seawater CO_2, and global temperature (Higgins and Schrag 2006; Panchuk et al. 2008; Zeebe et al. 2009). The first task is to estimate the CO_2 concentration of dissolved inorganic carbon (DIC is equal to $[CO_2] + [HCO_3^-] + [CO_3^{2-}]$) in the Paleocene ocean.

The starting point comes from the fact most C in the "mobile reservoirs" on the Earth's surface (ocean, atmosphere, and biosphere, including soils) is in the oceans—about 95% today—and most dissolved inorganic carbon in the oceans is present as HCO_3^-. The dissolved inorganic carbon concentration of the PETM oceans is unlikely to have differed much from today's value. The justification for this statement comes from constraints on the values of pCO_2, seawater calcium concentration, and the seawater $CaCO_3$ saturation during the PETM and today. By manipulating the equations describing equilibrium between dissolved CO_2, HCO_3^-, and CO_3^{2-} in seawater, and the solubility product of $CaCO_3$, one can derive the following equation:

$$[HCO_3^-] = (K_I \times K_{sp} / K_{II})^{1/2} \times ([CO_2] / [Ca^{2+}])^{1/2}.$$

The constants K_I and K_{II} are the dissociation constants of H_2CO_3 and HCO_3^-, and K_{sp} is the solubility product of $CaCO_3$. The chemical variable is the ratio of CO_2/Ca^{2+}. During the Paleocene and Eocene, it is thought that the atmospheric CO_2 concentration was higher than today by perhaps a factor of three to four (see chapter 6), while the seawater calcium concentration was higher by about a factor of two (Lowenstein et al. 2001). Because of the square root dependence on the CO_2/Ca^{2+} ratio, there would not have been a large change in the dissolved inorganic carbon concentration (DIC), which at seawater pH is approximately equal to $[HCO_3^-]$, despite large changes in ocean chemistry. The constant term would have been slightly lower than today (Millero 1979), because of its temperature dependence. Thus the best estimate is that the DIC inventory of seawater was perhaps 30% higher than the modern value of 40,000 Gt C. The difference, although uncertain, is not that large, and has been ignored in most studies.

Assuming that the DIC concentration of Paleocene seawater was similar to the modern concentration, we can estimate the magnitude of the CO_2 addition required to account for the observed $\delta^{13}C$ lowering at the onset of the PETM. The $\delta^{13}C$ of DIC is approximately zero, and the seawater DIC concentration was about 40,000 Gt C. If the carbon source was biogenic methane with a $\delta^{13}C$ approximately -60 ‰, then the required addition was $(-\Delta\delta^{13}C_{sw}/-60‰) \times 40,000$ Gt C. The $\Delta\delta^{13}C_{sw}$, the decrease in the seawater $\delta^{13}C$ value during the event, is itself an uncertain number, with estimates ranging from about 2 to 5‰ (e.g., Handley et al. 2008; Nunes and Norris 2006; Wing et al. 2005). The C input required for biogenic CH_4 is thus about 1300–3300 Gt C. If biogenic CO_2 was added instead of methane, the required input would have been about twice this amount.

Can carbon additions of this magnitude account for the observed warming and ocean acidification? At the lower levels of addition, probably not. While CH_4 is a powerful greenhouse gas, it is rapidly oxidized, and we therefore need to look to CO_2 to account for greenhouse warming over thousands of years. In one scenario, Zeebe et al. (2009) estimated that the addition of 3000 Gt C would raise CO_2 to 1700 ppm from an assumed initial value of 1000 ppm. If doubling CO_2 raises global temperature by 2–3°C, this change would account for a warming of only about one-third of the observed change of roughly 6°. Increasing the magnitude of the CO_2 release to 6000 Gt helps (Higgins and Schrag 2006; Panchuk et al. 2008). However, it may only be sufficient if Paleocene CO_2 is well below 1,000 ppm (in which case doubling CO_2 is easier). Alternatively, other feedbacks in the climate system may have contributed to the warming.

OCEAN ACIDIFICATION AND CACO$_3$ ACCUMULATION IN SEDIMENTS

Recent models also examine the quantitative effect of CO_2 addition on the magnitude of sedimentary $CaCO_3$ dissolution. To see how one approaches this problem it is necessary to describe the cycle of $CaCO_3$ in the oceans. Rivers carry dissolved Ca^{2+} and dissolved HCO_3^- to the oceans. As concentrations of these species rise in the sea, the solubility product of $CaCO_3$ will be exceeded. $CaCO_3$ will precipitate, and accumulate in sediments. The way most $CaCO_3$ precipitates out of seawater is as shells of Foraminifera and coccolithophorids. (Coccolithophorids are algae with external skeletons made up of intricate $CaCO_3$ plates, or liths.) When the organisms die, their skeletons fall to the seafloor. The flux of biogenic $CaCO_3$ to the seafloor is, today, about five times the river flux of Ca^{2+} and CO_3^{2-} to the oceans. Consequently, in the modern ocean, about 80% of sinking $CaCO_3$ dissolves. This fraction is dictated by feedback mechanisms that balance river input and sedimentation. If sediment accumulation were faster than the river input, for example, $CaCO_3$ would become less saturated in seawater, more of the $CaCO_3$ falling to the seafloor would dissolve, and a new balance would evolve.

Two effects cause deeper waters of the oceans to be undersaturated with $CaCO_3$, and surface waters to be supersaturated. First, organic matter forms in surface water and decomposes in deep water, according to the reaction:

$$CO_2 + H_2O \rightarrow CH_2O + O_2.$$

In deep water, respiration drives the opposing reaction. Production of organic carbon removes CO_2 from surface water, driving the equilibrium in this water toward higher CO_3^{2-} concentrations and higher degrees of $CaCO_3$ saturation. Respiration in the deep ocean has the opposite effect; $[CO_3^{2-}]$ decreases as it reacts with respiratory CO_2 to form HCO_3^-. Second, $CaCO_3$ has retrograde solubility. That is, it is actually more soluble in colder waters, and therefore has higher solubility in the deep sea. Thus, the solubility product increases with depth and the concentration product (observed $[Ca^{2+}] \times$ observed $[CO_3^{2-}]$) decreases with depth. The two curves cross at some intermediate depth. Above this depth, $CaCO_3$ is supersaturated in seawater and preserved in sediments; at deeper depths it is undersaturated and dissolves. The depth at which dissolution commences is the "lysocline" and the depth where the concentration of sedimentary $CaCO_3$ falls to zero is the "compensation depth." The Ca^{2+} and CO_3^{2-} concentrations of seawater adjust themselves, via feedback mechanisms, to levels where sedimentation of $CaCO_3$ is exactly in balance with input (fig. 7.3).

The addition of massive amounts of CO_2 to the oceans has the effect of lowering the CO_3^{2-} concentration of seawater via its reaction with CO_2 to make HCO_3^-. This process then lowers the degree of $CaCO_3$ saturation, promoting dissolution of sedimentary $CaCO_3$ in deep-sea

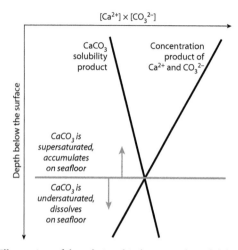

Fig. 7.3. Illustration of the relationship between the solubility product of CaCO$_3$ in the oceans and the concentration product. The Ca^{2+} concentration of seawater is nearly constant so that changes in the concentration product are due to changes in the concentration of carbonate ion.

sediments where CaCO$_3$ had previously accumulated. As a result CaCO$_3$ is present only at shallower depths than before the event. In the record, the compensation depth rises by a few hundred meters in the Pacific Ocean and well over a thousand meters in the Atlantic (see summaries in Panchuk et al. 2008; Zeebe et al. 2009). These authors show that one can account for the changes by invoking the addition of about 3000–6000 Gt C as CO$_2$, in the range of estimates from δ^{13}C.

POSSIBLE SOURCES OF CARBON DIOXIDE OR METHANE

Finally, there is the question of the source of CO_2 that initiated the event. Four possibilities have been suggested. Dickens et al. (1995) proposed that a warming of unknown cause initiated the release of CH_4 hydrates from deep-sea sediments. These solid-phase compounds form in sediments at moderate ocean depths when large quantities of CH_4 are produced from the decomposition of organic matter in the mud, and the temperature is cool enough that the stable form of CH_4 is as solid methane hydrates. Dickens et al. (1995) cite estimates that the modern CH_4 hydrate reservoir contains 8000–15,000 Gt C; the Paleocene reservoir would have been smaller because methane hydrates are less stable when ocean temperatures are warmer. A sudden warming of the oceans at the Paleocene-Eocene boundary could destabilize methane hydrates and cause CH_4 to be released to seawater. The observed $\delta^{13}C$ change at the onset of the PETM limits the resulting C flux to 1300–3300 Gt C, as estimated above. Recently Zeebe et al. (2009) worked backward, using the changes in carbonate compensation depth and seawater $\delta^{13}C$ to calculate the magnitude of the CO_2 release and its $\delta^{13}C$. They calculated a magnitude of 3000 Gt C and a $\delta^{13}C$ of $-50‰$. The very low $\delta^{13}C$ value, characteristic of methane, supports a large role for methane in the PETM event.

Higgins and Schrag (2006) suggested that the CO_2 came from oxidation of organic carbon in sediments of shallow water seas that dried up at the end of the Paleocene. In their hypothesis, large areas of shallow sea floor were uplifted and became emergent in regions of low rainfall. If the sedimentary organic carbon was converted to CO_2 by respiration or fires, these areas could easily supply the required amount. It is difficult to test this hypothesis; it would require identifying candidate areas and showing they lost large amounts of sedimentary carbon at the very beginning of the Eocene. Huber (2008) suggested that warming associated with the PETM may have essentially poisoned tropical plants by promoting photorespiration. Photorespiration is a process in which Rubisco, the enzyme that transforms CO_2 into organic C, also oxidizes organic C to CO_2. As temperatures rise, oxidation is favored at the expense of respiration, and when temperatures are sufficiently warm, plants cannot grow because of the oxidative losses. Death of tropical plants could conceivably lead to the release of ~1000 Gt C as CO_2 from both the plant and soil reservoirs. Against this hypothesis is the fact that CO_2 was elevated during the PETM. Higher CO_2 suppresses photorespiration and would have partially counteracted the increase of photorespiration caused by higher temperatures.

Then there is the intriguing coincidence of a massive carbon release at the start of the PETM, and massive volcanism in the North Atlantic associated with continental drift. The North Atlantic Igneous Province includes volcanic rocks extending from Greenland and Baffin Island at its west, to the Faroes, Great Britain, and northwestern Europe. It is associated with the onset of seafloor spreading in this region. There were extensive volcanic flows that coincided, within the uncertainty of a few hundred thousand years, with the onset of the PETM (Storey et al. 2007; Svensen et al. 2010; Westerhold et al. 2009). Some flows were injected as sills into layers of sedimentary rocks, and could have converted large amounts of organic carbon into methane. One can make back-of-the-envelope calculations showing that this process is a plausible source of the PETM releases, but there is not enough data to quantitatively estimate the associated carbon fluxes.

Finally, a recent paper offers the intriguing hypothesis that CO_2 released during the PETM was derived from the decay of organic matter in permafrost. Permafrost is permanently frozen ground. It contains relatively high levels of organic carbon derived from vegetation in the "active layer" (the surface soil in permafrost regions that defrosts in summertime). Warming leads to the deepening of the active layer, and respiration leads to the transformation of organic carbon (in the active layer) to CO_2 that is released to the atmosphere. DeConto et al. (2012) suggest that, during the early Cenozoic, Antarctica and Greenland were unglaciated, but largely covered with permafrost. Slow warming during the Paleocene (fig. 7.1) would have warmed large regions toward the point where permafrost would melt. Then, when changes in Earth's orbit around the sun favored warm summers (box 4), significant areas of permafrost would melt and release large amounts of CO_2 to the atmosphere. This release would lead to more warming, further release of

CO_2, more warming, and so forth, until the addition of CO_2 associated with the event was fully realized.

Whatever its origin, the PETM was not unique. There were at least two events, about 2 Myr after the PETM, that are similar in character but smaller in magnitude (Stap et al. 2010). These observations, like so much other data, give us a rich picture of the phenomenon, but it remains for us to put together the pieces and come up with the definitive picture of the origin of this event and the quantitative link between increased greenhouse gas concentrations and the observed warming.

REFERENCES

Papers with asterisks are suggested for further reading.

DeConto, R. M., S. Galeotti, M. Pagani, D. Tracy, K. Schaefer, T. Zhang, and D. Pollard (2012), Past extreme warming events linked to massive carbon release from thawing permafrost, *Nature*, 484, 87–92.

*Dickens, G. R., J. R. O'Neil, D. K. Rea, and R. M. Owen (1995), Dissociation of oceanic methane hydrate as a cause of the carbon isotope excursion at the end of the Paleocene, *Paleoceanography*, 10, 965–971.

Farley, K. A., and S. F. Eltgroth (2003), An alternative age model for the Paleocene-Eocene thermal maximum using extraterrestrial ³He, *Earth and Planetary Science Letters*, 208, 135–148.

Handley, L., P. N. Pearson, I. K. McMillan, and R. D. Pancost (2008), Large terrestrial and marine carbon and hydrogen isotope excursions in a new Paleocene/Eocene boundary section from Tanzania, *Earth and Planetary Science Letters*, *275*, 17–25.

Higgins, J. A., and D. P. Schrag (2006), Beyond methane: Towards a theory for the Paleocene-Eocene Thermal Maximum, *Earth and Planetary Science Letters*, *245*, 523–537.

Huber, M. (2008), A hotter greenhouse? *Science*, *321*, 353–354.

*Kennett, J. P., and L. D. Stott (1991), Abrupt deep-sea warming, palaeoceanographic changes and benthic extinctions at the end of the Palaeocene, *Nature*, *353*, 225–229.

Lowenstein, T., M. Timofeeff, S. Brennan, L. Hardie, and R. Demicco (2001), Oscillations in phanerozoic seawater chemistry: Evidence from fluid inclusions, *Science*, *294*, 1086–1088.

Millero, F. J. (1979), The thermodynamics of the carbonate system in seawater, *Geochimica et Cosmochimica Acta*, *43*, 1651–1661.

Nunes, F., and R. D. Norris (2006), Abrupt reversal in ocean overturning during the Palaeocene/Eocene warm period, *Nature Letters*, *439*, 60–63.

Panchuk, K., A. Ridgwell, and L. R. Kump (2008), Sedimentary response to Paleocene-Eocene Thermal Maximum carbon release: A model-data comparison, *Geology*, *36*, 315–318.

Rohl, U., T. Westerhold, T. J. Bralower, and J. C. Zachos (2007), On the duration of the Paleocene-Eocene Thermal Maximum (PETM), *Geochemistry, Geophysics, Geosystems*, *8*, 1–13.

Sluijs, A., S. Schouten, M. Pagani, M. Woltering, H. Brinkhuis, J. S. S. Damste, and G. R. Dickens (2006), Subtropical Arctic Ocean temperatures during the Palaeocene/Eocene Thermal Maximum, *Nature Letters*, *441*, 610–613.

Stap, L., L. J. Lourens, E. Thomas, A. Sluijs, S. Bohaty, and J. C. Zachos (2010), High-resolution deep-sea carbon and oxygen isotope records of Eocene thermal maximum 2 and H2, *Geology*, *38*, 607–610.

*Storey, M., R. A. Duncan, and C. C. Swisher, III (2007), Paleocene-Eocene Thermal Maximum and the opening of the northeast Atlantic, *Science*, *316*, 587–589.

Svensen, H., S. Planke, and F. Corfu (2010), Zircon dating ties NE Atlantic sill emplacement to initial Eocene global warming, *Journal of the Geological Society, London*, *167*, 433–436.

Weijers, J.W.H., S. Schouten, A. Sluijs, H. Brinkhuis, and J. S. S. Damste (2007), Warm Arctic continents during the Palaeocene-Eocene Thermal Maximum, *Earth and Planetary Science Letters*, *261*, 230–238.

Westerhold, T., U. Rohl, H. K. McCarren, and J. C. Zachos (2009), Latest on the absolute age of the Paleocene-Eocene Thermal Maximum (PETM): New insights from exact stratigraphic position of key ash layers +19 and −17, *Earth and Planetary Science Letters*, *287*, 412–419.

*Wing, S. L., G. J. Harrington, F. A. Smith, J. I. Bloch, D. M. Boyer, and K. H. Freeman (2005), Transient floral change and rapid global warming at the Paleocene-Eocene boundary, *Science*, *310*, 993–996.

*Zachos, J., G. Dickens, and R. Zeebe (2008), An Early Cenozoic perspective on greenhouse warming and carbon-cycle dynamics, *Nature*, *451*, 279–283.

Zachos, J., S. Bohaty, C. John, H. McCarren, D. Kelly, and T. Nielsen (2007), The Palaeocene-Eocene carbon isotope excursion: Constraints from individual shell planktonic Foraminifer records, *Philosophical Transactions of the Royal*

Society A-Mathematical, Physical and Engineering Sciences, *365*, 1829–1842.

Zachos, J., S. Schouten, S. Bohaty, T. Quattlebaum, A. Sluijs, H. Brinkhuis, S. Gibbs, and T. Bralower (2006), Extreme warming of mid-latitude coastal ocean during the Paleocene-Eocene Thermal Maximum: Inferences from TEX_{86} and isotope data, *Geology, 34*, 737–740.

Zeebe, R. E., J. C. Zachos, and G. R. Dickens (2009), Carbon dioxide forcing alone insufficient to explain Palaeocene-Eocene Thermal Maximum warming, *Nature Geoscience,* *2*, 576–580.

8 THE LONG COOLING OF THE CENOZOIC

THE OXYGEN ISOTOPE COMPOSITION OF BENTHIC FORA-
minifera provides a critical tool for characterizing the
cooling between the Paleogene and today, because it depends on temperature and ice volume. The first figure (fig.
6.1) of chapter 6 is an iconic graph of benthic foram $\delta^{18}O$
versus age from Zachos et al. (2001). It shows the major
climate events of the last 65 Ma. The dominant feature of
the climate record is the long cooling, starting around 50
Ma, that led to the ice age cycles of the Pleistocene, and to
our place in an interglacial period. This long-term trend
was punctuated by dramatic stepwise changes. By about 3
Ma, there was massive glaciation of the Northern Hemisphere continents at sea level, which will be discussed in
the next chapter.

The story here starts with the "ice-free" Earth in the
early Paleogene. In fact, the ice-free character of the
planet at this time is a matter of controversy, as discussed in chapter 6. We take the view that ice sheets
in the Paleogene were never close to being as extensive
as today. However, there may have been transient ice
sheets on Antarctica preceding the permanent glaciation beginning near the Eocene-Oligocene boundary
(34 Ma).

The $\delta^{18}O$ of benthic forams depends on ice volume and bottom water temperature. The role of the ice volume is to sequester isotopically light water in ice sheets, thereby making the residual ocean (and forams that grew in it) isotopically heavier (box 2). We can estimate the composition of the ice-free ocean by asking how much the isotopic composition of the oceans would change if today's ice sheets melted. The answer is that its $\delta^{18}O$ would fall by about 0.6‰. From this number, we can calculate that the temperature of the deep Paleogene ocean ranged from about 12°C (or perhaps slightly higher) during the Eocene climate optimum at 50 Ma, to about 5° immediately before the cooling at the Eocene-Oligocene boundary. For reference, the temperature of the modern deep ocean is about 2°C. There was thus a large change in the temperature of the deep ocean, and in the temperature of the high latitudes where deep waters form.

This long cooling was accompanied by a series of climate and biotic events that came to shape many features of the modern world. In this chapter we focus on four of these events: (1) the major coolings at 34 and 14 Ma, (2) the glaciation of Antarctica at 34 and 14 Ma, (3) the origin of the monsoons, and (4) the development of so-called C4 grasses in the later Miocene. We also examine a long period of stasis during the Oligocene. Other important changes transpired but are not discussed here. These include the warming from the Late Paleocene to the Eocene climate optimum, the long Eocene cooling from about 50 to 34 Ma, and the late Oligocene warming at about 25 Ma. We just don't know much about the nature of these events.

The scene is set in the previous two chapters. The Paleocene and Eocene periods were part of the long, warm interval from about 250 Ma to 34 Ma characterized as equable climates. Starting at the beginning of the Paleocene (65 Ma), the climate warmed until about 50 Ma, then cooled until a threshold was crossed, and Antarctica became permanently glaciated 34 million years ago.

THE COOLING AT THE EOCENE-OLIGOCENE BOUNDARY

The transition toward glacial conditions at the Eocene-Oligocene boundary was discovered by Kennett and Shackleton (1976) based on analyses of the $\delta^{18}O$ of benthic Foraminifera in a deep-sea drilling project core raised south of New Zealand (fig. 8.1). The $\delta^{18}O$ oscillates before the Eocene-Oligocene boundary, increases to a maximum, and then recovers toward slightly warmer values. The $\delta^{18}O$ increase reaches about +1.4‰ at its maximum. This change aggregates the effects of temperature and ice volume. Adopting a nominal value of 1‰ per 100 m of sea level for the ice volume dependence, and 0.25‰ per degree for the temperature dependence, we can explain the observed $\delta^{18}O$ change by glaciating Antarctica to its present extent (this involved ~54 m of sea level lowering) together with a cooling of 3–4°C. It seems unlikely that Antarctic ice volume was larger than present, but the cooling could have been larger and ice growth smaller.

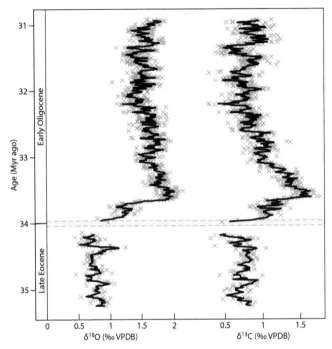

Fig. 8.1. The $\delta^{18}O$ and $\delta^{13}C$ of benthic Foraminifera (living on the sea floor) versus time across the Eocene-Oligocene boundary in the equatorial Pacific (ODP Site 1218, 8°53.38′ N, 135°22.0′ W) (Coxall et al. 2005). Increasing values of $\delta^{18}O$ (to the *right* in this diagram) indicate colder temperatures of ocean bottom waters and surface waters at high latitudes, together with increasing ice volume

The abundance of alkenones and tetraether compounds (box 3) in sediments across the Eocene-Oligocene boundary support this possibility of a larger temperature change and less ice growth (Z. Liu et al. 2009). The relative abundance of similar alkenones and tetraether compounds in plankton change as a function of surface water temperature. These temperature-dependent chemical variations are preserved by organic matter incorporated into deep-sea sediments. The relationship between abundance and temperature is known from the analyses of these compounds in surface sediments from the modern seafloor. From the abundance of alkenones and tetraether compounds, L. Liu et al. (2009) diagnosed a temperature decrease of about 5–7°C in seven deep-sea sediment cores from high latitudes.

The entire Eocene-Oligocene cooling and glaciation (denoted Oi-1) lasted about 400 Kyr (Coxall et al. 2005; Lear et al. 2008). The transition occurred in two roughly equal steps, with a ~200 Kyr plateau in the middle. After the second step, a warming began that erased about one-third of the $\delta^{18}O$ increase over the next one Myr. This warming notwithstanding, benthic foram $\delta^{18}O$ data clearly record a shift from a largely unglaciated Antarctic continent in the late Eocene to a heavily glaciated continent in the early Oligocene.

Glaciation at the Eocene-Oligocene boundary is accompanied by a deepening in the carbonate compensation depth, the deepest depth at which $CaCO_3$ skeletons are preserved in the sediments. Ocean chemistry thus changed in a way that was more favorable to $CaCO_3$ preservation, which is to say that the ocean became more alkaline. This change may have temporarily lowered the CO_2 partial pressure of surface ocean waters and hence the atmosphere (Coxall et al. 2005).

The growth of ice sheets in Antarctica and the cooling of the oceans, at least at high latitudes, were accompanied by major changes in continental temperatures, precipitation, vegetation, and animal life. In many areas, cool ecosystems representing open forest and grassland gradually replaced denser, warmer temperature forests (Bredenkamp et al. 2002; Jacobs et al. 1999). The change in flora drove a change in the mammalian fauna so extensive that it has been named the "Mongolian Remodeling" in Asia and the "Grande Coupure" in France.

In central North America, the oxygen isotope composition of fossil bones records a major cooling at the Eocene-Oligocene boundary (Zanazzi et al. 2007). As an air mass travels inland from its oceanic source, cooling induces condensation and precipitation. The precipitation is enriched in ^{18}O. By mass balance, residual water vapor, and subsequent precipitation, become progressively more depleted in ^{18}O along the travel path of the air mass. The colder the temperature at a point along the trajectory, the more water vapor will have been lost before the air mass reached that point, and the lighter will be the $\delta^{18}O$ of the precipitation. In this way, regional temperature is registered in the isotopic composition of precipitation. The isotopic composition of oxygen in bones at a fossil site reflects the isotopic composition of precipitation. Zanazzi et al. (2007) measured a $\delta^{18}O$ decrease across the Eocene-Oligocene boundary that reflected, according to their estimate, a cooling of about 8°C. Wolfe (1994) reported a similar cooling for western North America by examining the shapes of fossil leaves, which vary in a systematic way with temperature. In the Arctic, pollen studies indicate that wintertime temperatures fell by about 5°C. Seasonality increased, but Greenland was not heavily glaciated (Eldrett et al. 2009).

Blondel (2001) summarized evidence for a shift to more open vegetation across the Eocene-Oligocene boundary in western Europe. This shift indicates cooler temperatures, drier conditions, and increased seasonality. The vertebrate fauna changed simultaneously. Above the boundary, the diversity of indigenous ungulates decreased, ruminants became abundant, and dentition changed to reflect that animals began grazing on resistant plant material. Smaller animals became more important and larger animals declined. Changes in climate, flora, and fauna were similar in central Asia (Kraatz and Geisler 2010; Meng and McKenna 1998; Y.-Q. Wang et al. 2007). Less information is available about the Southern Hemisphere continents, but the general picture appears more or less the same (Bredenkamp et al. 2002; Jacobs et al. 1999).

As usual, there is some evidence that does not fit. For example, oxygen isotope data from Argentina and Germany show no evidence for a temperature change across the boundary (Heran et al. 2010; Kohn et al. 2004), and Prothero and Heaton (1996) found the North American fauna to be stable. Without doubt, in some areas changes in climate and vegetation were small. However, in most temperate and subpolar latitudes the changes were dramatic.

About eight Myr after the cooling and growth of ice at the Eocene-Oligocene boundary, there is a decrease in benthic foram $\delta^{18}O$ back toward values characteristic of the Late Eocene. Otherwise, there are no dramatic climate changes during the Oligocene. However, the benthic foram isotope record reveals persistent cyclicity in both $\delta^{18}O$ and $\delta^{13}C$ (Palike et al. 2006). Periods of variability fall into bands associated with changes in the eccentricity of Earth's orbit (100 and 400 Kyr), tilt of the spin axis (41 Kyr), and (to a small degree) precession (21 Kyr). The $\delta^{13}C$ varies primarily with a period of 400 Kyr, a reflection of higher frequency variations in the carbon cycle coupled to the long residence time of carbon in the oceans. Higher values of $\delta^{13}C$ probably reflect increased burial of organic carbon. Climate (indicated by $\delta^{18}O$) and carbon cycle variations are clearly coupled, and part of the story is that climates are cooler when more carbon is buried (and higher $\delta^{18}O$ coincides with higher $\delta^{13}C$ in benthic forams). Increased burial of organic C means less CO_2 in air. However, the small magnitude of $\delta^{13}C$ changes shows that CO_2 changes due to organic carbon burial alone would probably have been too small to affect climate. Such orbital-duration cycles pervade the entire Cenozoic climate record, although amplitudes, dominant periods, and the nature of links between $\delta^{13}C$ and $\delta^{18}O$ vary over time.

There was a transient cooling at the Oligocene-Miocene boundary (23.5 Ma), and a major Miocene cooling at about 14 Ma. In the context of the long Cenozoic record, this latter event marks the point of no return, leading to further cooling and eventually, intense glaciations of the Late Pleistocene.

THE MIDDLE MIOCENE COOLING (MI-1), 13.8 MA

At 13.8 Ma, the $\delta^{18}O$ of benthic forams rises by about 1‰ at many sites around the world (Flower and Kennett 1994; Holbourn et al. 2005; Savin et al. 1981; Zachos et al. 2001). This increase may be observed in the Cenozoic $\delta^{18}O$ curve of Zachos et al. (2011), (fig. 6.1), and in the more detailed record from Hole 1146 of the Integrated Ocean Drilling Program (IODP) from the South China Sea (fig. 8.2, Holbourn et al. [2005]). The Mg/Ca ratios of forams, which rise with temperature, show that most of the $\delta^{18}O$ increase was due to ice volume, indicating growth of an ice sheet on Antarctica comparable to today's. The sediment data is accompanied by field evidence for an expansion of Antarctic glaciers. This evidence is most compelling in the Dry Valleys region (near the coast south of New Zealand). Deposits prior to the $\delta^{18}O$ shift indicate quickly flowing mountain glaciers with water at the bottom, while after the shift they reflect large ice sheets frozen to the bedrock (Lewis et al. 2008).

The detailed record of $\delta^{18}O$ shows periodic variations beginning well before the cooling (fig. 8.2). The dating shows that, below the climate transition, individual climate cycles ($\delta^{18}O$ cycles) lasted about 40 Kyr, whereas above the transition, they lasted about 100 Kyr. These periods correspond, respectively, with periods of variations in the tilt of Earth's spin axis as well as variations in the eccentricity of Earth's orbit around the sun. Eccentricity itself leads to very small changes in annual radiation

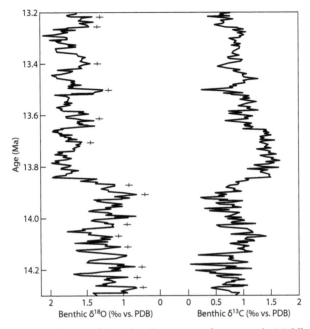

Fig. 8.2. The $\delta^{18}O$ and $\delta^{13}C$ in benthic Foraminifera across the Middle Miocene cooling and growth of the East Antarctic ice sheet at 13.9 Ma; these were collected at ODP Site 1146 (South China Sea). Plus signs mark warm times of climate oscillations recorded in the cores. From the "Supplementary Online Material" of Holbourn et al. (2005).

incident on the planet. However, it is important for climate change because, at periods of high eccentricity, a hemisphere will have very warm summers when it is positioned close to the sun during that season. The nature of changes in Earth's orbit around the sun, and the influences of these changes on solar insolation and climate, are outlined in box 4.

A remarkable feature of the record is the contrast between the symmetric nature of 40 Kyr obliquity cycles that precede the cooling, and the sawtooth nature of the 100 Kyr eccentricity cycles that follow. The warming and cooling phases of the tilt (obliquity) cycles appear about equal in length. Eccentricity cycles, on the other hand, involved long, slow cooling, followed by a rapid warming. As we will see, this pattern was repeated over the past ~1 Myr or so, and has received much attention. The 40 Kyr cycles may be explained as a direct response of the climate system to summer insolation (Huybers and Tziperman 2008). The 100 Kyr cycles are more complex. Their dynamics may be similar to the 100 Kyr cycles of the Late Pleistocene, discussed in the next chapter.

CAUSES OF OLIGOCENE AND MIOCENE COOLINGS

Declining concentrations of atmospheric CO_2 are obvious candidates as the culprits of Cenozoic glaciation. The most complete proxy record for Cenozoic CO_2 comes from Pagani et al. (2005), (fig. 8.3), who analyzed $\delta^{13}C$ in alkenones from deep-sea sediments. When the CO_2 concentration of air is higher, oceanic phytoplankton produce organic carbon more depleted in ^{13}C. Alkenones are long-chained carbon compounds that are particularly stable, and thus desirable for reconstructing $\delta^{13}C$ in ancient organic matter. Converting alkenone $\delta^{13}C$ to CO_2 concentrations requires accounting for a variety of effects influencing growth rate and the availability of CO_2 for photosynthesis

Box 4
Orbital Variations and Climate

One of the main attributes of glacial-interglacial climate change is its cyclical nature. While debate continues about the origin of the 100 Kyr cycle observed in the ice ages of the last 500–1000 Ka, it is clear that much periodicity in climate is caused by changes in Earth's orbit around the sun. There are three properties of Earth's orbit that vary with time. The first is the tilt (obliquity) of the spin axis, which changes from about 21.5° to 24.5° during a period of 41 Kyr. The second is the eccentricity, which changes with several periods, including 100 Kyr and 400 Kyr. The third is the wobble in the tilt axis (precession), which changes with a period of about 21 Kyr (actually 19 and 23). The nature of precession is such that, if at a certain point on its orbit, the spin axis points toward the sun, then 10.5 Kyr later it will be pointing away, and 21 Kyr later it will once again point toward the sun at the same orbital position. The properties are illustrated in figure A.

According to the eponymous theory of Milankovitch, the great continental ice sheets will retreat when northern hemisphere summers are warm, and advance when northern hemisphere summers are cold. The focus is on the northern hemisphere because this is where we find temperate and sub-polar lands that host glaciers during ice ages, and because these areas are ice free during interglacial times. In the southern hemisphere, temperate and subpolar latitudes are almost entirely ocean, and great ice sheets cannot develop north of Antarctica. Milankovitch focused on summer, because he thought that this was the critical season for maintaining ice sheets. His idea was that, if summers were cool, ice sheets could be preserved to resume their growth in the fall. If summers were warm, however, ice sheets would melt and earth would tend toward an interglacial climate.

(Box 4 continued)

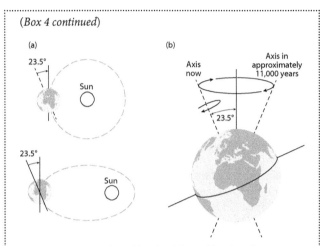

Fig. A. Properties of Earth's orbit (*a*) Earth's orbit when eccentricity is low (*left, top*) and when it is high (*left, bottom*). In this example, it is summer in the southern hemisphere, which is tilted toward the sun. Summer is cooler under the high eccentricity condition. (*b*) The precession of the spin axis and the role of obliquity. Tilt is 23.5° in this diagram (as today), which is an intermediate tilt condition. When tilt increases to 24.5°, summers in both hemispheres will be warmer, and when it decreases to 21.5°, summers will be cooler.

Northern hemisphere summers are cool under three conditions. The first occurs when tilt is small. With a lower tilt, high latitudes are oriented more obliquely to the sun, and summers are cooler. When tilt is large, high latitudes are more inclined toward the sun in summers, which will then be warmer. The second condition for cool summers is a high eccentricity. This is the condition under which the planet is both closest to and furthest from the sun at different points during one year. When eccentricity is high, northern hemisphere summers will

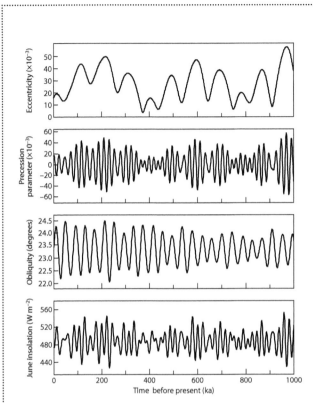

Fig. B. Insolation properties as a function of time for the last 1000 Kyr (back to 1 Ma). From top to bottom, eccentricity, precession parameter (a function of precession and eccentricity indicating the distance of Earth from the sun at northern hemisphere summer solstice), obliquity (tilt of the spin axis), and June insolation at 65° N. Insolation is maximum when obliquity is high and the precession parameter is low (i.e., when Earth is close to sun at northern summer solstice).

(Box 4 continued)

sometimes occur when Earth is furthest from the sun, a favorable condition for the growth of ice sheets. The third condition for cool summers concerns the precession of Earth's spin access. Northern hemisphere summers will be cool when the axis precesses to the point where summertime actually falls at that point along the orbit where Earth is furthest from the sun.

Precession, variations in eccentricity, and variations in obliquity lead to an irregular pattern of northern hemisphere summer insolation with time (fig. B). Imprinted in this pattern are periods of 21, 41, 100, and 400 Kyr that derive from the periods of precession, tilt, and eccentricity. These same frequencies are embedded in many climate records, proving that orbital changes pace glacial-interglacial climate change (and climate change during unglaciated periods as well). Furthermore, absolute dating confirms Milankovitch's prediction that increasing summertime insolation melts ice sheets, and decreasing summertime insolation enables their growth.

in phytoplankton. Consequently, derived atmospheric CO_2 concentrations have large uncertainties. However, there seem to be three robust results in figure 8.3. First, CO_2 concentrations were much higher in the Eocene than in more recent times. Second, there was a large CO_2 decrease around the time of the Eocene-Oligocene boundary. Third, CO_2 concentrations probably didn't vary much during the Miocene, although absolute values might have been higher than shown in this reconstruction.

Falling CO_2 is therefore an attractive explanation for the growth of ice on Antarctica at the Eocene-Oligocene

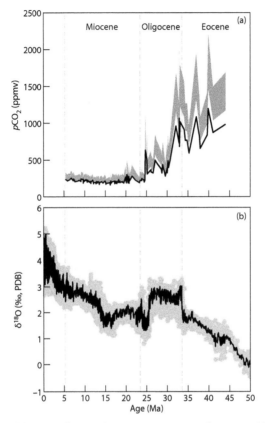

Fig. 8.3. (*a*) Atmospheric CO_2 concentration over the past 50 Myr reconstructed from the analysis of the $\delta^{13}C$ of alkenones in deep-sea sediments. The thin black line represents the minimum estimated CO_2 concentration. (*b*) The $\delta^{18}O$ of benthic Foraminifera separated from deep-sea sediments. Higher values indicate larger continental ice sheets and colder temperatures. From Pagani et al. 2005.

boundary. Detailed studies of alkenone $\delta^{13}C$ across the Eocene-Oligocene boundary give strong evidence, from many locations, for a CO_2 drop at about this time (Pagani et al. 2011). A modeling study by DeConto and Pollard (2003) outlines how the glaciers might have developed. According to their simulation, mountain glaciers would have existed in the Late Eocene when, in their model, the CO_2 concentration was 840 ppm (3 times the pre-industrial value). Then with falling CO_2, glaciers would advance to lower elevations, and their growth would be enhanced by the ice albedo. Eventually the snowline would descend to the elevation of the East Antarctic Plateau, whereupon much of Antarctica would rapidly become glaciated. According to their model, Antarctica would be largely ice covered with 560 ppm CO_2 in the atmosphere.

The alkenone $\delta^{13}C$ proxy record does not give any evidence for a decrease in CO_2 at ~14 Ma (Pagani et al. 2005; Shevenell et al. 2004). On the other hand, Holbourn et al. (2005) speculate that CO_2 did in fact decline, based on two arguments. First, the $\delta^{13}C$ of seawater reached a maximum at this time, which may reflect the enhanced burial of organic carbon. Organic carbon preferentially removes ^{12}C, and burial of organic C thus leaves the oceans enriched in the heavy isotope. Burial of organic carbon would lower atmospheric pCO_2. Second, the lowering of sea level would have led to less burial of $CaCO_3$ on continental shelves and more in the deep ocean basin, depressing the $CaCO_3$ compensation depth and making the oceans more alkaline.

The cooling and ice growth at 13.8 Ma was followed by a small progressive cooling over the next 10 Ma or so. During this interval, and indeed throughout the Cenozoic, vegetation was changing in response to climate and co-evolving with herbivores. The signature change, to which we turn next, was the spread of grasslands at the expense of forests and the appearance of so-called C4 plants.

CHANGING CLIMATES AND THE RISE OF C4 GRASSES DURING THE CENOZOIC

The distribution of vegetation on the planet near the end of the Cretaceous, based on the distribution of fossil plants, is shown in figure 6.3 (Upchurch et al. 1998). Vegetation differed from that on the modern Earth in three ways. First, high-latitude areas that are currently covered by ice sheets or tundra were then forested, reflecting the equable climates discussed earlier. Second, some arid regions where deserts prevail today were then also forested, reflecting a wetter planet. Third, all Cretaceous vegetation fixed CO_2 into organic carbon by what is known as the C3 pathway, whereas today about 75% of photosynthesis is by the C3 pathway and 25% is by the C4 pathway (these are described below). These changes in vegetation are in turn linked to a profound change in the nature of animals, with which plants evolved.

The history of grasslands begins with the evolution of grasses in the Cretaceous and Paleocene. Then, in the Oligocene and Miocene, forests in areas of Asia, Africa, North America, and South America began thinning out, and grasslands became more common. By approximately 15 Ma, grasslands were extensive on Asia, Africa, and North and South America (Jacobs et al. 1999). One factor likely contributing to the change from woodlands to grasslands was the development of seasonally arid climates. These allow for fires in the dry season, which destroy trees and favor grasses.

The change in vegetation was accompanied by a shift in herbivores from browsers to grazers. Groups of animals adapted to changing vegetation during the Cenozoic in more specific ways. For example, horses and plants engaged in an arms race in which horse teeth became progressively longer with time, while grasses evolved that were increasingly hard to chew.

In the ancient C3 metabolic pathway, the enzyme Rubisco transforms CO_2 into an organic carbon compound with three carbon atoms (hence the name C3). This transformation is costly in that Rubisco also oxidizes organic C to CO_2, and today about 30% of fixed carbon is lost by this process of "photorespiration." Plants utilizing the C4 pathway evolved during the Oligocene or perhaps earlier (Edwards et al. 2010). Photosynthesis using the C4 pathway avoids photorespiratory losses by loosely attaching a CO_2 molecule to a C3 compound to make a C4 compound, and then transporting the C4 compound to the isolated site of Rubisco. The C4 compound then releases its CO_2 molecule, providing a high-CO_2 environment around Rubisco in which photorespiratory losses are suppressed. Of course, this process has its own energetic costs. Thus, whether C3 or

C4 plants are favored depends on the relative losses due to photorespiration and the additional effort associated with C4 photosynthesis. Photorespiration is more rapid at higher temperatures and lower CO_2 concentrations. Falling CO_2 therefore favors C4 plants. On the other hand, cooling temperatures and wet climates favor C3. When conditions are conducive to forestation, C3 plants dominate because there are no C4 trees.

While plant fossils are rare, we have a detailed isotopic record of the rise of C4 plants because they discriminate differently against the heavy carbon isotope. The $\delta^{13}C$ of C3 plants is about $-27‰$, whereas that of C4 plants is about $-12‰$. The $\delta^{13}C$ of an animal's food is registered in the $\delta^{13}C$ of carbonate fluorapatite, the principle mineral in teeth. In precipitating this mineral, plants enrich the ^{13}C by about 14‰. Thus, the $\delta^{13}C$ of CO_3^{2-} in teeth—the best preserved skeletal component—is about $-13‰$ for herbivores eating C3 plants, and about $+1‰$ for herbivores eating C4. By measuring $\delta^{13}C$ in fossil teeth, one can determine the food source and, inferentially, the abundance of C3 and C4 plants on the landscape (Cerling et al. 1997).

In Asia, Africa, and North and South America, abundant data show that there was a shift from C3 to C4 flora starting at 8–9 Ma and completed, depending on location, over the next few million years (fig. 8.4). This shift is observed in teeth studied in the seminal work of Thure Cerling and colleagues (Cerling et al. 1997). It is also found in $CaCO_3$ formed in soils (Badgley et al. 2008) that acquired most of their CO_2 from decaying plant material. The shift in flora was likely associated with an

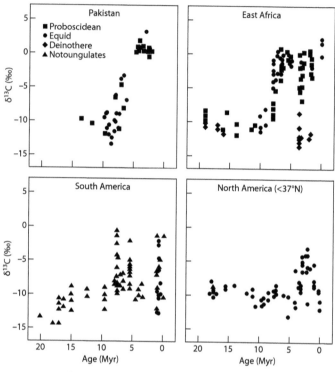

Fig. 8.4. The $\delta^{13}C$ of tooth enamel from fossil teeth over the past 20 Myr from 4 geographic regions (Cerling et al. 1997). The shift toward heavier (less negative) $\delta^{13}C$ at about 9 Ma reflects a shift in the nature of the metabolic pathways in grasses from C3 to C4.

extension in the area of grasslands, and was clearly connected with dramatic faunal changes. For example, Badgely et al. (2008) show that in Pakistan, many browsers and fruit-eating animals died out across the transition, at about 9 Ma.

The cause(s) of the transition from C3 to C4 grasses is a matter of controversy. Photosynthesis in C4 grasses is an adaptation that evolved independently, starting at least as long ago as the early Oligocene (34 Ma) in many different plant species (Edwards et al. 2010). Its development would be favored by lower CO_2 concentrations in air. The $\delta^{13}C$ history of alkenones in deep-sea sediments do not indicate a fall of CO_2 at about 9 Ma (fig. 8.3) when the big increase in tooth $\delta^{13}C$ occurs. But Kurschner et al. (2008), among others, argue that CO_2 was much more variable during the Miocene than the carbon isotope data indicate. Other suggestions are that C4 plants expanded because of increasingly arid conditions (e.g., L. Liu et al. 2009), or because they developed modes of resistance to grazers. Several of these factors probably contributed.

ORIGIN OF THE INDIAN AND EAST ASIAN MONSOONS

Strong summer monsoons derive from heating over land during the summer months. The oceans warm also, but less so, because vertical mixing distributes the heat increase. Heating over land leads to rising air, which in turn leads to precipitation. Uplift also sucks in wet air from the adjacent ocean, maintaining the moisture supply. Airflow is reversed in winter, with air sinking over land. The land is arid, and becomes a source of dust.

The climates of India and East Asia are strongly influenced by monsoons. These events have an ancient origin, extending back to at least the Eocene, according to

Huber and Goldner (2012). The nature of precipitation since that time has been strongly affected by the uplift of the Himalayas and the formation of the Tibetan Plateau, which began by at least 35 Ma (Molnar et al. 2010) as a consequence of the collision between the northward-drifting Indian subcontinent and Asia. Uplift increased precipitation to the south of the Himalayas and led to a rain shadow, thus increasing aridification to the north and northeast (Huber and Goldner 2012; Molnar et al. 2010). By 24 Ma, dust from desert regions was transported to what is now the Chinese Loess Plateau by the westerlies and the northwesterly winds of the winter monsoon (Qiang et al. 2011; Sun et al. 2010). Around 8 Ma, aridification became more intense and deposition of loess on the plateau grew more extensive (Liu et al. 2009; Molnar et al. 2010). The Indian and East Asian monsoons were fully active.

Thus, by the Miocene-Pliocene boundary (at 5.3 Ma), Earth had been transformed from a wetter planet with equable climates throughout, including extensive forests and moderate climates at high latitudes, to a cooler, drier planet with a massive permanent ice sheet on the Antarctic continent. Large continental landmasses were covered by desert or grasslands, and recently evolved C4 plants dominated the landscape in warmer areas. Fauna coevolved with flora, leading to the extinction of many browsers and the appearance of diverse assemblages of grazing animals. The Himalayas had risen, deserts covered much of central Asia, and highly seasonal monsoonal climates were in place in India and East Asia.

These aggregate changes set the stage for increasingly intense cyclic glaciations on the Northern Hemisphere continents, the hallmark of Pliocene and Pleistocene climates.

REFERENCES

Papers with asterisks are suggested for further reading.

Badgley, C., J. C. Barry, M. E. Morgan, S. V. Nelson, A. K. Behrensmeyer, T. E. Cerling, and D. Pilbeam (2008), Ecological changes in Miocene mammalian record show impact of prolonged climatic forcing, *Proceedings of the National Academy of Sciences of the United States*, 105, 12145–12149.

Blondel, C. (2001), The Eocene-Oligocene ungulates from western Europe and their environment, *Palaeogeography, Palaeoclimatology, Palaeoecology*, 168, 125–139.

Bredenkamp, G. J., F. Spada, and E. Kazmierczak (2002), On the origin of Northern and Southern Hemisphere grasslands, *Plant Ecology*, 163, 209–229.

*Cerling, T. E., J. M. Harris, B. J. MacFadden, M. G. Leakey, J. Quade, V. Eisenmann, and J. R. Ehleringer (1997), Global vegetation change through the Miocene/Pliocene boundary, *Nature*, 389, 153–158.

*Coxall, H. K., P. A. Wilson, H. Palike, C. H. Lear, and J. Backman (2005), Rapid stepwise onset of Antarctic glaciation and deeper calcite compensation in the Pacific Ocean, *Nature*, 433, 53–57.

*Deconto, R. M., and D. Pollard (2003), Rapid Cenozoic glaciation of Antarctica induced by declining atmospheric CO_2, *Nature*, 21, 245–249.

*Edwards, E. J., C. P. Osborne, C.A.E. Stromberg, S. A. Smith, and C. G. Consortium (2010), The origins of C_4 grasslands: Integrating evolutionary and ecosystem science, *Science*, *328*, 587–591.

Eldrett, J. S., D. R. Greenwood, H.I.C., and M. Huber (2009), Increased seasonality through the Eocene to Oligocene transition in northern high latitudes, *Nature*, *459*, 969–974.

Flower, B., and J. Kennett (1994), The middle Miocene climatic transition: East Antarctic ice sheet development, deep ocean circulation and global carbon cycling, *Palaeogeography, Palaeoclimatology, Palaeoecology*, *108*, 537–555.

Heran, M.-A., C. Lecuyer, and S. Legendre (2010), Cenozoic long-term terrestrial climatic evolution in Germany tracked $\delta^{18}O$ of rodent tooth phosphate, *Palaeogeography, Palaeoclimatology, Palaeoecology*, *285*, 331–342.

Holbourn, A., W. Kuhnt, M. Schulz, and H. Erlenkeuser (2005), Impacts of orbital forcing and atmospheric carbon dioxide on Miocene ice-sheet expansion, *Nature*, *438*, 483–487.

Huber, M., and A. Goldner (2012), Eocene monsoons, *Journal of Asian Earth Sciences*, *44*, 3–23.

Huybers, P., and E. Tziperman (2008), Integrated summer insolation forcing and 40,000-year glacial cycles: The perspective from an ice-sheet/energy-balance model, *Paleoceanography*, *23*. doi: 10.1029/2007PA001463.

Jacobs, B. F., J. D. Kingston, and L. L. Jacobs (1999), The origin of grass-dominated ecosystems, *Annals of the Missouri Botanical Garden*, *86*, 590–643.

*Kennett, J. P., and N. J. Shackleton (1976), Oxygen isotopic evidence for the development of the psychrosphere 38 Myr ago, *Nature*, *260*, 513–515.

Kohn, M. J., J. A. Josef, R. Madden, R. Kay, G. Vucetich, and A. A. Carlini (2004), Climate stability across the Eocene-Oligocene transition, southern Argentina, *Geology*, *32*, 621–624.

Kraatz, B. P., and J. H. Geisler (2010), Eocene-Oligocene transition in Central Asia and its effects on mammalian evolution, *Geology*, *38*, 111–114.

Kurschner, W. M., Z. Kvacek, and D. L. Dilcher (2008), The impact of Miocene atmospheric carbon dioxide fluctuations on climate and the evolution of terrestrial ecosystems, *Proceedings of the National Academy of Sciences of the United States*, *105*, 449–453.

*Lear, C. H., T. R. Bailey, P. N. Pearson, H. K. Coxall, and Y. Rosenthal (2008), Cooling and ice growth across the Eocene-Oligocene transition, *Geology*, *36*, 251–254.

Lewis, A., D. R. Marchant, A. C. Ashworth, L. Hedenas, S. R. Hemming, J. V. Johnson, M.J. Lent, et al. (2008), Mid-Miocene cooling and the extinction of tundra in continental Antactica, *Proceedings of the National Academy of Sciences of the United States*, *105*, 10, 676–10, 680.

Liu, L., J. T. Eronen, and M. Fortelius (2009), Significant mid-latitude aridity in the Middle Miocene of East Asia, *Palaeogeography, Palaeoclimatology, Palaeoecology*, *279*, 201–206.

Liu, Z., M. Pagani, D. Zinniker, R. DeConto, M. Huber, H. Brinkhuis, S. R. Shah, et al. (2009), Global cooling during the Eocene-Oligocene climate transition, *Science*, *323*, 1187–1190.

Meng, J., and M. C. McKenna (1998), Faunal turnovers of Palaeogene mammals from the Monogolian Plateau, *Nature*, *394*, 364–367.

Molnar, P., W. R. Boos, and D. S. Battisti (2010), Orographic controls on the climate and paleoclimate of Asia: Thermal and mechanical roles for the Tibetan Plateau, *Annual Reviews of Earth and Planetary Science*, *38*, 77–102.

*Pagani, M., J. Zachos, K. Freeman, B. Tipple, and S. Bohaty (2005), Marked decline in atmospheric carbon dioxide concentrations during the Paleogene, *Science*, *309*, 600–603.

Pagani, M., M. Huber, Z. Liu, S. Bohaty, J. Henderiks, W. Sijp, S. Krishnan, and R. DeConto (2011), The role of carbon dioxide during the onset of Antarctic glaciation, *Science*, *334*, 1261–1264.

Palike, H., R. D. Norris, J. O. Herrle, P. A. Wilson, H. K. Coxall, C. H. Lear, N. J. Shackleton, et al. (2006), The heartbeat of the Oligocene climate system, *Science*, *314*, 1894–1898.

Prothero, D. R., and T. H. Heaton (1996), Faunal stability during the Early Oligocene climatic crash, *Palaeogeography, Palaeoclimatology, Palaeoecology*, *127*, 257–283.

Qiang, X., Z. S. An, Y. G. Song, H. Chang, Y. B. Sun, W. G. Liu, H. Ao, et al. (2011), New Eolian red clay seqeunce on the western Chinese Loess Plateau linked to onset of Asian desertification about 25 Ma ago, *Science China Earth Sciences*, *54*, 136–144.

Savin, S., R. Douglas, G. Keller, J. Killingley, L. Shaughnessy, M. Sommer, E. Vincent, and F. Woodruff (1981), Miocene benthic formainiferal isotope records: A synthesis, *Marine Micropaleontology*, *6*, 423–450.

Shevenell, A., J. Kennett, and D. Lea (2004), Middle Miocene Southern Ocean cooling and Antarctic cryosphere expansion, *Science*, *305*, 1766–1770.

Sun, J., J. Yie, W. Wu, X. Ni, S. Bi, Z. Zhang, W. Liu, and J. Meng (2010), Late Oligocene-Miocene mid-latitude aridification and wind patterns in the Asian interior, *Geology*, *38*, 515–518.

Upchurch, G., B. Otto-Bliesner, C. Scotese, T. Lenton, P. Valdes, and D. Beerling (1998), Vegetation-atmosphere interactions and their role in global warming during the latest Cretaceous, *Philosophical Transactions of the Royal Society B- Biological Sciences*, *353*, 97–112.

Wang, Y.-Q., J. Meng, X. Ni, and C. Li (2007), Major events of Palaeogene mammal radiation in China, *Geological Journal*, *42*, 415–430.

Wolfe, J. A. (1994), Tertiary climatic changes at middle latitudes of western North America, *Palaeogeography, Palaeoclimatology, Palaeoecology*, *108*, 195–205.

Zachos, J., M. Pagani, L. Sloan, E. Thomas, and K. Billups (2001), Trends, rhythms, and aberrations in global climate 65 Ma to present, *Science*, *292*, 686–693.

Zanazzi, A., M. J. Kohn, B. J. MacFadden, and D. O. Terry Jr. (2007), Large temperature drop across the Eocene-Oligocene transition in Central America, *Nature*, *445*, 639–642.

9 THE ORIGIN OF NORTHERN HEMISPHERE GLACIATION AND THE PLEISTOCENE ICE AGES

THE ICE AGES OF THE PLEISTOCENE—THE PAST 2.6 MIL-
lion years—are among the most compelling events in the
history of Earth's climate. These were periods in which
Earth was remarkably transformed. During the peaks of
the ice ages, glaciers covered the northern part of Eurasia
and the northern and central parts of North America. In
the eastern United States, their southern extent reached
almost to my home in central New Jersey. Away from the
ice, most continental areas were significantly colder and
much drier, and the vegetation was correspondingly im-
poverished. The oceans cooled as well, wintertime sea ice
extended further equatorward, and there were important
changes in patterns of ocean circulation. Two key factors maintained these colder climates: (1) the ice sheets
reflected more of sun's heat back to space, and (2) greenhouse gas concentrations were lower. During the Pleistocene, these two factors worked in tandem (providing
positive feedback to one another); in other words, lower
CO_2 led to ice sheet growth, greater albedo, and colder
climates, which in turn led to changes in ocean circulation that further lowered atmospheric CO_2, and so on.

The origin of glacial climates extends back at least 34 Myr, when a large ice sheet grew on Antarctica. At about 3 Ma, Earth crossed a climate threshold as ice sheets began to grow and decay on the Northern Hemisphere continents. Climate has varied since that time in a manner forced by changes in Earth's tilt, precession, and eccentricity. While climate has responded to orbital forcing during much or most of Earth history, the amplitude of the responses and the detail in which they are recorded are unique to this period. Over the past 5 Ma and beyond, the warmest periods of each cycle have tended to be slightly cooler, the coldest times have become much colder, the amplitudes of the cycles have increased, and their period has switched from 41 Kyr to about 100 Kyr. We are presently living in the warm phase of the current 100 Kyr glacial cycle.

We begin this chapter with a description of the planet during the height of the last ice age (~20 Ka). We then backtrack to the early Pliocene (~ 5 Ma), and discuss the warmth of this period. Returning toward the present, we discuss in turn: the origin of Northern Hemisphere glaciation, at about 3 Ma; the "40K (obliquity-dominated) world" between 2.5–1 Ma; the "100K world" of the past million years; and orbitally driven climate changes at low latitudes.

EARTH'S SURFACE DURING THE LAST ICE AGE

Nine continental ice sheets formed the hallmark feature of the glacial Earth (fig. 9.1). The footprints of the West

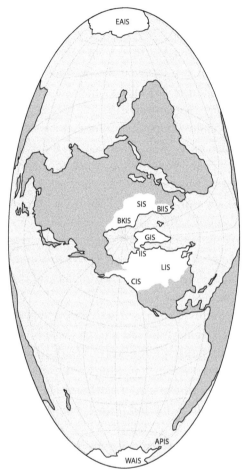

Fig. 9.1. Boundaries of the major ice sheets of the last glacial maxi-
mum, or LGM. EAIS, East Antarctic ice sheet; SIS, Scandinavian ice
sheet; BIIS, British Isles ice sheet; BKIS, Barents-Kara ice sheet; GIS,
Greenland ice sheet; ILS, Inuitian ice sheet; LIS, Laurentide ice sheet;
CIS, Cordilleran ice sheet; APIS, Antarctic Peninsula ice sheet; WAIS,
West Antarctic ice sheet. From Clark et al. (2009).

Antarctic ice sheet and the East Antarctic ice sheet were similar to today's, but larger. The East Antarctic ice sheet was grounded in the Ross Sea, and the West Antarctic ice sheet extended further than present over the continental shelf (Conway et al. 1999; Hall et al. 2000; Peltier 2004; Peltier and Fairbanks 2006; Polyak et al. 2008). The glacial Greenland ice sheet also extended out over the continental shelf (Dyke 1999), rather than ending at the edge of present Greenland. Through an ice bridge crossing Baffin Island, it was contiguous with the Laurentide ice sheet (centered over Hudson's Bay) and ultimately with the Cordilleran ice sheet, which reached to the Pacific (Peltier 2004). In Eurasia, the British Isles ice sheet covered most of Great Britain and, to the west, extended onto the continental shelf (Ballantyne 2010). To the east, the British Isles ice sheet was continguous with the Scandinavian and the Barents-Kara ice sheets. The Barents-Kara sheet hugged the northern coast of Siberia, was grounded in what is now the Barents and Kara Seas, and extended north all the way to Svalbard (Peltier 2004; Polyak et al. 2008).

In the tropics and midlatitudes, snowlines (lowest elevation extents) of mountain glaciers were much lower than today. These "equilibrium line altitudes" (ELA's) descended by about 1000 m, although with considerable variability from one location to another (Mark et al. 2005). Equilibrium line altitudes are determined largely by summer temperatures since ice melts when summertime temperatures rise above the freezing point. A snowline descent of 1000 m corresponds to a cooling of about 6°C or more.

The last major change in the cryosphere concerns the extent of sea ice. In the North Atlantic, reconstructions based on ice-sensitive microfossils indicate a band of wintertime sea ice covered the basin north of a line extending from the west of Ireland to Iceland, and to Newfoundland (de Vernal et al. 2005; Pflaumann et al. 2003). In the Atlantic sector of the Southern Ocean, wintertime sea ice extent was about twice that at present, with ice extending 5–10° north of its present position (Gersonde et al. 2005; Wolff et al. 2006).

With CLIMAP (1976), a program to map ocean temperatures during the last ice age, and later programs, great effort has gone into mapping sea surface temperatures using a variety of proxies (box 3). The most recent synthesis of these data comes from the multiproxy approach for the reconstruction of the glacial surface ocean (MARGO) project (MARGO 2009). There was cooling during the last ice age of about 2–3°C in the eastern equatorial Pacific and Atlantic Oceans (Dubois et al. 2009; Herbert et al. 2010; Lawrence et al. 2006; Lea et al. 2000), western equatorial Pacific (Herbert et al. 2010; Lea et al. 2000; Stott et al. 2002; Visser et al. 2003), eastern equatorial Atlantic (Herbert et al. 2010), and western equatorial Indian Ocean (Herbert et al. 2010). Stronger winds and more intense upwelling may have contributed to this cooling.

The central gyres of the ocean appear to have cooled very slightly. Poleward of 45°, the cooling was greater: the eastern North Atlantic cooled by 8°, and the eastern North Pacific cooled by a similar amount. South of 60° S, the Southern Ocean cooled in most areas by 3–10° (Gersonde et al. 2005; MARGO 2009), a possible consequence of shifts in the positions of boundaries of the major water masses. There is actually some evidence for warming of Arctic waters between Greenland and the British Isles (de Vernal et al. 2005; Pflaumann et al. 2003).

Continental temperatures have been reconstructed primarily based on the distribution of fossil pollen, which reflects the local vegetation. The dominant feature of these studies is that, almost the world over, vegetation became sparser; desert areas expanded, some forests were replaced by grasslands, and closed forests were replaced by open forests (Harrison and Prentice 2003; Marchant et al. 2009; Prentice et al. 2000). These changes in the flora resulted from three environmental shifts: temperatures grew cooler, precipitation lessened, and the CO_2 concentration of air decreased (e.g., Harrison and Prentice 2003).

In the midlatitudes, cooling was quite significant. Estimates for summertime cooling in Eurasia range from about 5 to 10°C, and estimates for wintertime cooling range from about 12 to 20°C (Jost et al. 2005; Wu et al. 2007). There is considerable uncertainty in these estimates, and considerable variability from place to place. Temperature changes in the tropics were smaller, on the order of 3 to 5°C. Values toward the higher end of the range are supported by studies of noble gas concentrations in groundwater (e.g., Stute et al. 1995). The temperature signal in this proxy comes from the fact that solubilities of dissolved gases increase at lower temperatures. One can therefore sample groundwater that dates back to the last glacial maximum (LGM), measure noble gas concentrations, and calculate the temperature required to produce the observed values.

The pollen data indicate an increase in aridity nearly worldwide. Precipitation, estimated from pollen data by Wu et al. (2007), show decreases in Eurasia averaging anywhere from 10 to 40 cm/year, depending on the method used for the reconstruction. Forests were almost nonexistent in Eurasian midlatitudes during the LGM. Steppe was replaced by a similar biome but with more drought-tolerant plants, and forests were replaced by grasslands (Harrison and Prentice 2003; Prentice et al. 2000). Further north, climates at the edge of the glaciers were very cold and dry, and tundra dominated the surface. Forests in North America and the tropics were more open than today (Colinveaux et al. 2000; Harrison and Prentice 2003; Wu et al. 2007). Along the southwestern part of South America, cold rainforest was displaced by grass and shrubland (Marchant et al. 2009). This shift may reflect a decrease in the strength of the westerly winds (Rojas et al. 2009), illustrating the fact that some regional changes in precipitation were induced by meteorological shifts rather than general alterations in the global climate.

In some areas, meteorological conditions changed in a way that led to increased precipitation. Perhaps the best example is the Great Basin, in the western United States. Here steppe was replaced by open forests at the LGM (Prentice et al. 2000), and the region hosted lakes that were far larger than their modern counterparts (Broecker and Orr 1958; Broecker and Kaufman 1965). This change has been attributed to a southern movement in the position of jet stream (e.g., Braconnot et al. 2007).

GLACIAL-INTERGLACIAL CHANGES IN ATMOSPHERIC GREENHOUSE GAS CONCENTRATIONS

Any attempt to understand the causes of the ice ages must take into account variations in the concentrations of greenhouse gases. A major breakthrough came around 1980 when French and Swiss groups independently published CO_2 records based on the analysis of gases trapped in ice cores; these showed that glacial CO_2 values were significantly lower than the preindustrial level (Delmas et al. 1980; Neftel et al. 1982). (The nature of ice core climate records, and the significance of the trapped gases, is discussed in box 5.) With time, ice core scientists learned how to best extract and analyze the trapped gases, and the precision of CO_2 measurements improved. One record of CO_2 and CH_4 changes during the last glacial maximum and the termination of the ice age is shown in figure 9.2. This record shows CO_2 rising 80ppm from its glacial value of about 185 ppm to 265 ppm at 10 Ka. The CO_2 concentration continued to rise during the Holocene and reached the preindustrial value of 280 ppm. Also shown in this figure is the CH_4 rise from 350 to nearly 700 ppb with deglaciation. Given estimates for climate sensitivity, these CO_2 and CH_4 changes

Box 5
Ice Cores

Ice cores have yielded records of polar climates and allowed us to reconstruct histories of atmospheric greenhouse gases. They preserve a wealth of information with a variety of proxies. The three most important climate properties gleaned from ice core studies are polar temperature history, atmospheric greenhouse gas history, and information about the aridity of source areas for dust.

Glaciers form in polar areas as snow falls and accumulates. As a glacier grows, the weight at the center begins squeezing ice out the side, much as the pressure of your fingers squeezes toothpaste out of the tube. If the climate is stable, a glacier will eventually reach a steady state, in which accumulation is balanced by loss. The thickness of a year's snow accumulation at the surface must be identical to the precipitation rate. At depth, however, the annual layers are thinner, because a fraction of ice has been squeezed out by lateral flow; the weight of ice pushing downward squeezes deeper ice out the sides. In principle, the thickness of annual layers will decrease to the glacier bed, but the age of ice at the bed will correspond to the time the region first became glaciated. In practice, there is shear as one approaches the bed, introducing folding that scrambles the ice stratigraphically, and allows old ice to be transported out of the system. This scrambling of ice limits the age of the oldest ice that can be easily sampled. In polar cores drilled to date, the oldest ice in the continuous record is 123 Ka in Greenland and 800 Ka in Antarctica. Older ice is present, but not in continuous stratigraphic order.

The primary record in ice cores is the isotopic temperature of the ice, either in terms of $\delta^{18}O$ or δD, where δD represents that ratio of $^2H/^1H$ in ice (see box 1). The proxy quality of the isotopes is as follows: When water evaporates from the ocean, the light isotope evaporates faster. As a water mass travels to colder areas, gas periodically condenses out to form rain or snow. The condensation process involves isotopic equilibrium. The heavy isotope is enriched in precipitation, leaving the residual water vapor depleted in the heavy isotope. As the air mass travels to colder and colder regions, it loses more and more precipitation, and the heavy isotope depletion in the residual water vapor becomes greater and greater. By the time air gets to East Antarctica, water gas is depleted in the heavy oxygen isotope by about 6% (60‰), and in the heavy isotope of hydrogen by about 40% (400‰)! Quantitatively, the depletion of the heavy isotopes depends on the fraction of water vapor initially present that is lost to condensation when the air reaches East Antarctica. This loss is a function of the temperature dependence of the saturation water vapor concentration in air. Therefore, the heavy isotope depletion ($\delta^{18}O$ or δD) depends on the temperature change between the source of evaporation and the site of precipitation. Since polar temperatures have changed far more than the lower latitude ocean temperatures where precipitation originates, the isotopic change of the ice gives a measure of temperature change in the polar region.

Ice cores contain salts and dust that give information about climate change. Arguably the most important ionic impurity for climate reconstruction is the sea salt concentration (or sodium concentration), which gives a measure of sea surface roughness and wind speed. Particulate impurities, including $CaCO_3$, originate as windblown dust from dry areas of the continents. Changes in their abundance reflect increasing or decreasing aridity in the continental source areas.

The most distinctive feature of the ice core record is the inclusion of fossil air. Snow falling at the surface is actually composed of small ice crystals with a porosity of about 65%.

(*Box 5 continued*)
As a layer of snow is buried by subsequent snowfall, it recrystallizes such that the individual crystals become larger and the porosity decreases. By a depth of 50–100 m, the porosity has decreased to about 10%, and the snow grains are packed together so closely that they completely enclose individual bubbles of air. Beyond this point, the ice is impermeable, and there will be no significant separation of ice and its trapped gases. We can now extract the trapped gases by either melting or crushing the ice, and characterize the chemistry of the ancient atmosphere. In this way, ice core scientists have reconstructed the atmospheric history of CO_2, CH_4, and N_2O, among other gases, as well as their isotopic composition.

would lead to a global temperature increase of about 2°C. The climate forcing from CH_4 is about 20% that for CO_2.

Understanding the ice core CO_2 record starts with noting long-term CO_2 feedbacks, in which Earth's temperature adjusts to the level where CO_2 uptake by weathering

Fig. 9.2. Climate records spanning the last glacial termination, from 20 to 10 Ka (from Broecker et al. 2010). (*a*) Isotopic temperature in East Antarctica (EPICA Dome C core; upward on the graph indicates warmer temperatures). (*b*) Atmospheric CO_2 concentration. (*c*) Isotopic temperature of Greenland (GISP2 core; upward on the graph indicates warmer temperatures). (*d*) The $\delta^{18}O$ of $CaCO_3$ from cave deposits in East Asia (lighter values are thought to reflect stronger summer monsoons). (*e*) Atmospheric CH_4 concentration, from GISP2. Gray bands mark times of warming in Antarctica and relatively cold times in Greenland.

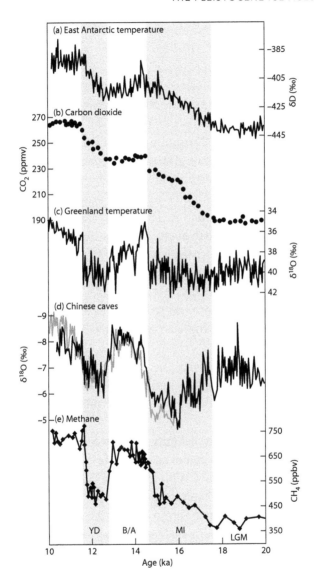

balances the input to the oceans and atmosphere from outgassing (chapter 4). These processes fixed the average atmospheric CO_2 concentration at about 230 ppm over the past 800 Kyr. Over shorter timescales, atmospheric CO_2 can vary as CO_2 is transferred between Earth's mobile reservoirs (air, ocean, and plants/soils). In his classic paper, Broecker (1982) summarized the major processes leading to glacial-interglacial CO_2 variations, and outlined how their effects could be read in the sedimentary record. His starting point was to determine that of all the CO_2 in Earth's mobile reservoirs, about 1–2% is in the atmosphere, 5% in plants and soils, and ~94% in the oceans. Since the oceans contain the lion's share of the CO_2, explaining atmospheric variations requires looking primarily at changes in the oceans.

We have now come to recognize at least six factors that contribute to the glacial-interglacial difference in atmospheric CO_2:

1. *Ocean temperature.* The lower temperatures of the ice age oceans caused an increase in the solubility of CO_2 in seawater, and contributed to lowering the CO_2 concentration of air.

2. *Ocean salinity.* Salinity increased during the ice ages, as water was removed from the oceans to make continental ice sheets. With less water in the oceans, the concentration of dissolved inorganic carbon rose, leading to an increase in the atmospheric CO_2 concentration.

3. *Mass of carbon in the land biosphere and soil carbon reservoirs.* During the ice ages, glaciers covered areas currently occupied by forests and taiga. The biomass carbon

and soil carbon were oxidized to CO_2, raising the CO_2 inventory of the atmosphere and ocean, and increasing the CO_2 concentration of air.

4. *Strength of the biological pump.* The "biological pump" is the term used to describe the fixation of organic matter in the sunlit surface ocean, its sinking, and its oxidation at depth (chapter 1). This process does not change the total inventory of biologically active elements (C, N, P, etc.) in the oceans, but it influences the distribution. The biological pump causes the concentrations of bioactive elements to be low in the surface ocean and high in the deep ocean. By lowering the concentration of DIC in the surface ocean, it also lowers the partial pressure of CO_2 in the surface ocean and the concentration of CO_2 in air. Evidence for a stronger biological pump during the ice ages suggests that this change contributed to decreasing CO_2 in the glacial atmosphere.

5. *Residence time of water in the deep ocean.* As a consequence of the biological pump, bioactive elements are stripped from surface water and added to deep water. The fluxes are reversed by ocean mixing; deep water coming to the surface has high concentrations of dissolved inorganic carbon and nutrients, while surface waters mixing to depth have low concentrations. Evidence suggests that mixing was more sluggish during glacial times. As a consequence, more DIC would have been sequestered in the deep sea. The surface water pCO_2, and consequently the atmospheric CO_2 concentration, would have dropped.

6. *Changes in deep-sea calcium carbonate saturation.* During glacial times, the deep ocean had a higher

burden of metabolic CO_2, and was more acidic. This increase in acidity led to the dissolution of forams and coccoliths that make up the bulk of deep sea sediments over much of the ocean. Dissolution of $CaCO_3$ raises the concentration of CO_3^{2-}, which reacts with dissolved CO_2 to make bicarbonate. The concentration of dissolved CO_2 decreases in the process. This decreased CO_2 concentration is transmitted to the surface by mixing, and lowers the CO_2 concentration of surface waters.

The challenge, then, is to understand how these factors acted during the ice age, and combined to suppress the atmospheric CO_2 concentration by about 80 ppm compared to its typical interglacial value. According to recent estimates of Kohler et al. (2005) and Sigman and Boyle (2000), the glacial temperature decrease (factor 1, above) lowers pCO_2 by ~30 ppm, the salinity increase (factor 2) raises CO_2 by 7 to 12 ppm, and the glacial diminution of forests and soil carbon (factor 3) raises glacial CO_2 by 15 to 34 ppm. The CO_2 changes from these three processes nearly cancel, and we need to look to factors 4, 5, and 6 (above) in order to explain the glacial CO_2 lowering of ~80 ppm.

In fact there is evidence that each of these three processes has played a significant role. We start with the idea that a stronger biological pump contributed to lower glacial CO_2. Broecker (1982) originally invoked a higher oceanic phosphate concentration to increase the strength of the biological pump during glacial times. This increase in phosphorus, along with a nitrate increase due to nitrogen fixation, would lead to enhanced growth of phytoplankton during the ice age and the transfer of CO_2 from surface to deep waters by the biological pump. Subsequently, several authors pointed out that the Southern Ocean is the region where changes in the biological pump would have the greatest impact on atmospheric CO_2 (e.g., Sarmiento and Toggweiler 1984). The basis for this idea is that much of nitrate and phosphate in Southern Ocean surface waters goes unused. Rather than being taken up by organisms that will eventually sink, thereby depleting DIC in surface waters, these nutrients simply remain as dissolved species, descending to depth in the dissolved form as new deep waters form. These "unutilized nutrients" thus make no contribution to DIC drawdown and the associated lowering of surface water pCO_2. This situation contrasts with the other region of deep water formation, the North Atlantic, where nutrients are almost completely drawn down, thus already making the maximum possible contribution to atmospheric CO_2 decreases.

In this context, one can imagine two ways of enhancing the strength of the biological pump and lowering oceanic and atmospheric pCO_2 during the ice age in the process. One is to increase the export of organic carbon to depth in the Southern Ocean. Martin (1990) proposed that this happened as a result of increased ice age aridity and stronger winds, which would increase the transport of iron-bearing dust to the Southern Ocean. Here, phytoplankton have plenty of nitrate and phosphate but are starved for iron, which they require for photosynthesis. More iron would relieve this limitation, increasing the strength of the biological pump and lowering atmospheric CO_2. A second means of enhancing the biological pump is to increase stratification (Sigman et al. 2010). The fertility of ecosystems depends, in part, on the availability of light in the wind-mixed surface layer, which extends to tens of meters depth in summer. For a given flux of solar radiation to the ocean surface, the average flux of light in the mixed layer will decrease as the mixed layer becomes deeper. Sigman et al. proposed that mixed layers were shallower during the ice ages, leading to higher productivity within that layer, a greater drawdown of DIC and nutrients, and lower CO_2 both in the surface waters of the Southern Ocean and the atmosphere.

Slowing the exchange between CO_2-rich deep waters of the ocean interior and the surface waters of the Southern Ocean (factor 5) also likely contributed to lower ice age atmospheric CO_2 concentrations. Slowing the exchange would cause the deep ocean to retain more CO_2, and would suppress the release to the atmosphere of metabolic CO_2 that had accumulated in the deep ocean (Sigman et al. 2010). Sigman et al. (2010) also proposed a single explanation to account for both increased stratification and the slower mixing of Southern Ocean surface waters during glacial times. They suggested that, during the ice ages, westerly winds that are today centered over the Southern Ocean shifted to the north. Weaker winds over the Southern Ocean would lead to both shallower mixed layers and a slower rate of exchange between regional surface waters and the ocean interior.

In summary, colder glacial ocean temperatures lead to a sizable decrease in atmospheric CO_2 (factor 1), but this decrease is roughly balanced by the effect of higher ocean salinity (factor 2) and the transfer of carbon from the land biosphere and soils to the oceans (factor 3). A large part of the CO_2 decrease is associated with a strengthening of the biological pump (factor 4) and a slowing of ocean circulation, such that the deep ocean retains more CO_2 (factor 5). Factors 4 and 5 together lead to dissolution of $CaCO_3$ on the seafloor, and increased deep ocean alkalinity. Mixing imprints thus decreased alkalinity on the surface waters, further lowering the CO_2 concentration of air.

WHY WAS EARTH COLDER DURING THE ICE AGES?

Earth's temperature is controlled by the radiative balance as outlined in chapter 1. Our planet's orbit around the sun is changing continuously, and these changes influence times at which different latitudes receive their sunlight. However, the total amount of sunlight Earth receives in one trip around the sun is nearly constant. Instead, three other factors accounted for the cooling of the ice age Earth (e.g., Henrot et al. 2009; Jahn et al. 2005). First, the presence of large ice sheets increased the planetary albedo (reflectance), accounting for a cooling of about 3°C. Second, the switch to less abundant vegetation also increased the planetary albedo (deserts reflect more sunlight than forests), contributing a cooling of about 1°C.

Finally, the lower concentration of greenhouse gases, notably CO_2, led to a further 2° cooling. These three factors together account for the cooling of about 6°C during the ice ages relative to interglacial times.

EVOLUTION OF EARTH'S CLIMATE SYSTEM OVER THE PAST FIVE MILLION YEARS

We now turn back the clock and discuss the transitions to cooler climates over the past 5 Ma, leading to the glacial periods described above. The plot of $\delta^{18}O$ of benthic Foraminifera versus time (fig. 9.3) represents much of the variability in Earth's climate system over the past 5.4 Ma. As discussed in box 2 and chapter 8, changes in the $\delta^{18}O$ of benthic forams record changes in bottom water temperature and global ice volume. For reference, a 1‰ increase in $\delta^{18}O$ indicates growth of ice sheets that lower sea level by about 100 m, cooling of about 4°C, or some combination of the two. The plot in figure 9.3 is constructed by averaging, or "stacking," individual records of the benthic foram $\delta^{18}O$ from 57 deep-sea sediment cores. It represents global climate change during the past 5.4 Ma, with higher $\delta^{18}O$ values indicating colder temperatures and more ice.

The data in this figure record four distinct climate periods separated by three major transitions. The warmest interval of the past 5 Ma comes at the beginning, in the early Pliocene. At this time, Earth was about 3°C warmer than today, the amplitude of glacial-interglacial cycles was modest, and ice volume may have been about half

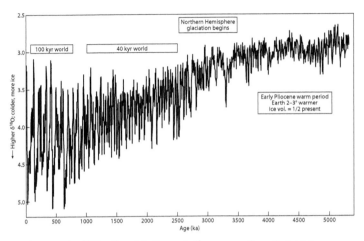

Fig. 9.3. The $\delta^{18}O$ of benthic Foraminifera versus time, 0–5.4 Ma. This curve was derived by "stacking," or averaging, records from 57 deep-sea sediment cores (Lisiecki and Raymo 2005).

that at present. There is no evidence for large ice sheets in the Northern Hemisphere, except for Greenland. Then, beginning at about 3.5 Ma, extreme $\delta^{18}O$ values of the cycles became greater (colder) than in the Early Pliocene and the amplitude of the cycles began to increase, recording a transition period ending in periodic ice ages with large glaciers in the Northern Hemisphere. The first of the large cycles began at about 2.8 Ma. From this time until about 1.0 Ma, there were large climate cycles of about 40 Kyr duration. This era is known as the "40 K world." Around 1 Ma, there was another transition, leading to climate cycles that lasted about 100 Kyr, which doubled the amplitude of the 40 Kyr cycles. Earlier, we discussed the coldest part of the most recent cycle, the last glacial maximum. The termination of this ice age led to the Holocene (see the next chapter), a relatively long period of warm climate that hosted the evolution of civilization.

These observations raise the following five questions, which we will discuss in turn:

1. Why was the Early Pliocene world warmer than today's?
2. What caused the initiation of glaciation on North America and Eurasia about three million years ago?
3. Why was climate cycling during the 40 K world controlled by Earth's obliquity (axial tilt) to the exclusion of precession?
4. What caused the transition, about 1 Ma, from the 40 K world to the longer and larger climate cycles of the 100 K world?
5. What dynamics explain the 100 Kyr duration of climate cycles over the past 1 Myr?

The Warmer World of the Early Pliocene

Evidence for warmer conditions in the Pliocene comes from coastal deposits indicating a higher contemporaneous sea level (and consequently smaller ice volumes than today), pollen studies indicating more equable climates on land, and proxy data from deep-sea sediments recording higher water temperatures both in the surface and on the seafloor. The most extensive summaries, which form the basis of the following discussion, pertain to a time interval centered at around 3 Ma, and hence may already capture a small amount of the Pliocene cooling associated with the beginning of glaciation in North America and Eurasia.

Evidence for a higher sea level comes from Pliocene shorelines found well above modern sea level in tectonically stable coastal areas, in the sequence of deposition and erosion of coral reef atolls, and in the lower $\delta^{18}O$ of benthic foraminifera in deep-sea sediment cores (Dowsett, Robinson, Stoll, and Foley 2010). These results indicated that sea level was higher by about 25 m, with an uncertainty of perhaps $\pm$ 10 m. This sea level change would correspond, for example, with much smaller ice sheets in Greenland and West Antarctica, and an East Antarctic ice sheet about 70% of the present size.

Ocean temperatures were, on average, 2–3° C warmer than today. Changes in two regions were particularly significant. First, the eastern equatorial Pacific was nearly as warm as the west (Dowsett, Robinson, Haywood, et al. 2010; Wara et al. 2005), leading to the suggestion that the region was in a "permanent El Niño" mode. While this view is controversial, clearly the mean state was closer to El Niño conditions than it is today. Second, high-latitude warming relative to the modern climate was somewhat greater than low-latitude warming (Dowsett, Robinson, Haywood, et al. 2010). High-latitude warming was particularly strong in the North Atlantic, where temperatures between Greenland and Europe rose by 5°C or more (Dowsett, Robinson, Haywood et al. 2010).

Conditions on land are known from extensive mapping of pollen deposits. The reconstruction by Salzmann et al. (2008) is based on extrapolating pollen observations as guided by modeling. In general climates were warmer and wetter than today. Subtropical deserts of Africa, Australia, and Asia were more restricted, and were partly replaced by dry shrub biomes. Southern Africa was somewhat wetter and more heavily vegetated than today. In northern latitudes, temperate forests displaced taiga, and taiga displaced tundra.

There are two leading explanations for the warmer conditions of the early- to mid-Pliocene. The first is that the atmospheric CO_2 concentration was higher than today. The second is that warmer conditions in the tropics led to warmer climates in temperate and polar regions. Both explanations find considerable support.

Evidence for higher CO_2 in the Pliocene comes from two proxy approaches. The first, as for earlier times in the Cenozoic, is the $\delta^{13}C$ of selected long-chain hydrocarbons (alkenones). As discussed above, phytoplankton prefer to assimilate ^{12}C; the extent to which they actually exclude ^{13}C depends on the availability of dissolved CO_2 in ocean surface water and hence in the atmosphere. The second proxy is the isotopic composition of boron (B) in biogenic $CaCO_3$. The isotopic composition of boron in $CaCO_3$ registers the pH of the parent seawater. Given an estimate of the dissolved inorganic carbon concentration, one can then estimate the partial pressure of CO_2 in seawater, and hence the CO_2 concentration of air.

Two recent papers use these approaches to reconstruct atmospheric pCO_2 back to the Pliocene and beyond (Pagani et al. 2009; Seki et al. 2010). Both present proxy data supporting higher CO_2 concentrations in the Pliocene, amounting to about 400 ppm compared to preindustrial values of 280 ppm. Modeling studies show that CO_2 at this level could be responsible for the difference between modern and Pliocene climates (e.g., Vizcaino et al. 2010). Thus, higher CO_2 is a strong candidate for causing a warm Pliocene, but we should remain cautious. While most records indicate a higher CO_2 concentration, the records are not coherent (changes in different locations occur at different times), there is a high uncertainty and a lot of scatter in reconstructed CO_2 concentrations, and CO_2 levels reverse in some cases as one goes further back into the still warmer climates of the Miocene. The $\delta^{13}C$ of alkenones is a good proxy for identifying large CO_2 changes, such as those between the Eocene and the present, but $\delta^{13}C$ and boron isotopes are more challenging to apply to the characterization of smaller changes of the past 5 Ma.

The second hypothesis for a warmer Pliocene—that warmer tropics led to warmer climates in the high latitudes—may seem unlikely. However, a role for the tropics in ice age climate has long been advocated by tropical climatologists, because most of Earth's heat is captured within 30° of the equator. The analysis of modern temperature variability shows that boreal regions of North America and Eurasia are warmer during times of El Niño events. In fact, modern trends indicate that these boreal regions would be ice free in the presence of a permanent Pliocene El Niño (Huybers and Molnar 2007; Vizcaino et al. 2010). Fedorov et al. 2006 outlined one mechanism linking boreal climates to the tropical Pacific; they suggested that clear skies with high water vapor concentrations would replace highly reflective stratus clouds currently overlying cool surface waters in the eastern equatorial Pacific. Albedo would decrease and the greenhouse effect would increase, thereby warming the planet.

The attribution of Plio-Pleistocene cooling to the onset of colder conditions in the tropics begs the question of why the tropics cooled during the Pliocene. One idea is that high-latitude cooling caused the thermocline in tropical waters to rise to shallower and shallower depths. The thermocline is the depth interval in which the temperature of subsurface waters cools from the warm surface value to lower temperatures characteristic of the intermediate and deep ocean. At first, surface waters would have continued to be warm, with cooler subsurface waters found closer and closer to the surface. Eventually, cooler waters would have come all the way to the surface when winds were favorable (i.e., during non El Niño times), as they do today (Fedorov et al. 2006). Alternatively, seafloor spreading and continental drift has changed the exact position of islands in the western equatorial Pacific. New Guinea and Australia have drifted to the north, and the island of Halmahera has grown. These changes may have redirected cool North Pacific waters into the Indian Ocean, cooling that basin (Cane and Molnar 2001; Karas et al. 2009). Again this cooling would have contributed to the growth of ice sheets in boreal regions.

Origin of Northern Hemisphere Glaciation

The onset of glaciation in different Northern Hemisphere regions generally cannot be documented from field deposits (moraines, glacial outwash, etc.), because these "near field" deposits would have been eroded by subsequent glaciers. However, early glaciers that reached the coast, and calved into the ocean, did leave permanent records in the form of ice-rafted detritus accumulating in deep-sea sediments. "Ice-rafted detritus" consists of grains of continental rocks that are entrained into glacial ice, enter the sea in calving icebergs, and are released to fall to the seafloor when the icebergs melt. They are functionally similar to dropstones, but are much more widely distributed because of their small size, and can be sampled in seafloor sediments collected by coring. Their composition may be diagnostic of specific source regions.

According to evidence from ice-rafted detritus, glaciers were sometimes present on Greenland as far back as the Late Miocene. Greenland glaciation then became more extensive at about 3.3 Ma (Kleiven et al. 2002; St. John and Krissek 2002). At about 2.74 Ma, the abundance and footprint of ice-rafted detritus expanded, with Scandinavian and North American sources leaving their mark in the North Atlantic and North Pacific Oceans, respectively (Haug et al. 2005; Jansen et al. 2000; Shackleton et al. 1984; St. John and Krissek 2002).

As in other contexts, the benthic foram oxygen isotope record (fig. 9.3) provides an integrated index of the extent of glaciation. Clearly, an increase in the intensity of glacial conditions occurred sometime after 3.5 Ma. A detailed analysis of the individual records suggests a gradual intensification between 3.6 and 2.4 Ma (Mudelsee and Raymo 2005). This intensification was followed by a stationary period about 300 Kyr in duration, after which cooling resumed. At 2.74 Ma, the first of a series of large $\delta^{18}O$ cycles began. These repeated, every 40 Kyr, until about 1 Ma. The $\delta^{18}O$ data thus seem to record the expansion of ice sheets on Antarctica, Greenland, Eurasia, and North America, culminating with the development of ice sheets on Eurasia that reached the shoreline and calved into the sea by 2.74 Ma.

We are not lacking for explanations of the origin of Northern Hemisphere glaciation, although none provides a fully satisfying answer. Explanations fall into four groups. First, tectonic changes led to changes in ocean circulation and meteorology that in turn led to glaciation. Second, the shoaling of the equatorial Pacific thermocline, and the end of the permanent El Niño, cooled the Northern Hemisphere continents and in turn induced glaciation. Third, the CO_2 concentration fell to the point where glaciation was possible. Fourth, the hydrography of the North Pacific Ocean changed in a way that led to elevated rates of snowfall over North America.

Consider the first hypothesis—tectonic changes. A number of Pliocene tectonic events have been invoked to help explain the onset of Northern Hemisphere glaciation. Perhaps the most robust is the northward drift of New Guinea that was mentioned earlier. By ~3–4 Ma, this motion would have lead to the flow of cooler North Pacific water into the Indian Ocean (Cane and Molnar 2001), and perhaps a shallower thermocline (cool waters closer to the surface) in the equatorial Pacific as well (Karas et al. 2009). Other possible tectonic influences on ocean circulation and climate include the uplift of Central America, which cut off a seaway connecting the Caribbean to the eastern equatorial Pacific; the final uplift of the Himalayas and mountains in the American West; and changes in the bathymetry of the Greenland-Iceland-Norwegian Seas (Maslin et al. 1998; Molnar 2008). It is unclear whether these events would have led to significant cooling, and even the direction of their influence on climate is uncertain.

The second hypothesis for the origin of Northern Hemisphere glaciation involves the end of "permanent El Niño" conditions in the eastern equatorial Pacific. Earlier, we discussed evidence that this change is partly responsible for the difference between Pliocene and Pleistocene climates. However, the divergence in temperatures between the eastern and western equatorial Pacific occurs at different times in different records, depending on what season and water depth are recorded by the proxies. The times do not necessarily correspond to major climate events in the benthic foram record. For example, eastern and western Pacific temperature trends in some records diverge at about 2.2 Ma, well after the onset of Northern Hemisphere glaciation (Etourneau et al. 2010; Ravelo 2010; Wara et al. 2005). We thus need to look elsewhere to understand the origin of northern glaciation, or the increased amplitude of glacial cycles at 2.7 Ma.

The third hypothesis for the onset of Northern Hemisphere glaciation is a decrease in the atmospheric CO_2 concentration to a low level that would enable the growth of ice sheets. As discussed earlier, proxy data that is based on both boron isotopes and the $\delta^{13}C$ of CO_2, suggest such a change during the past 5 Ma. However, the various records do not yet provide the coherent picture of timing that one would need to assess the role of CO_2 in specific climate events.

The fourth hypothesis involves the paradoxical observation, involving proxy studies, that late summertime temperatures rose in the North Pacific at 2.7 Ma, even while wintertime temperatures cooled (Haug et al. 2005). The late summertime warming is attributed to enhanced stratification, which slowed mixing and kept summertime heat in the shallow mixed layer. Haug et al. suggest that the summertime warming would have extended into autumn. Warm autumns would increase North Pacific evaporation and hence snowfall over North America, accelerating the growth of ice sheets.

Finally, the question arises as to why the amplitude of sustained glacial-interglacial cycles increased at 2.74

Ma, rather than somewhat earlier or later. Maslin et al. (1998) noted that the onset of glaciation approximately corresponds to an increase in the amplitude of insolation changes associated with tilt, eccentricity, and precession. Until about 1 Ma, glacial-interglacial cycles correspond to the period of tilt (41 Kyr), making this parameter the key to understanding the timing of Pliocene glaciation. The amplitude of tilt cycles changes from about 1.4° at ~2.75 Ma, to 1.8° at ~2.6 Ma, to 2.5° at ~2.5 Ma. The amplitude of precession cycles increases markedly from 2.8 to 2.7 Ma. The argument follows that changes in the position of landmasses and ocean gateways, decreases in CO_2, and cooling in the eastern equatorial Pacific all combined in some way to enable the ice ages. Large amplitude glacial-interglacial cycles would subsequently commence as soon as orbital conditions were favorable, with alternating periods of especially cool summers in which Northern Hemisphere ice sheets would grow, and warm summers in which they would melt. (The nature of changes in Earth's orbit around the sun, and its role in glacial-interglacial climate change, are outlined in box 4.) This argument remains highly speculative but finds some support in recent modeling studies (Vizcaino et al. 2010).

The "40K World," 2.7–1 Ma

Between 2.5 and 1 Ma, climate cycles appear to have had a duration of ~40 kyr, based on both simple visual analysis (fig. 9.3) and statistical studies (Martinez-Garcia et al. 2010). The period of 41 Kyr is identical to the period of variations in Earth's tilt, and strongly suggests that glacial-interglacial cycles were simple responses to this orbital property. What is surprising, however, is that in the benthic foram $\delta^{18}O$ record of this time interval there is very little signature of changes in eccentricity and precession, which also lead to large insolation changes. Two hypotheses address this conundrum.

Raymo et al. (2006) noted the phase differences associated with tilt and precession. Increased tilt leads to warmer summers in both hemispheres. High precession forcing, on the other hand, leads to warmer summers in the hemisphere that enjoys summer close to the sun, and cooler summers in the hemisphere that endures summers far from the sun. Raymo et al. (2006) proposed that, when tilt is low (cool summers), ice sheets grow simultaneously in the Northern and Southern Hemispheres, producing a 41 Kyr signature in the $\delta^{18}O$ record. When precession forcing is strong, however, growth of ice sheets in one hemisphere accompanies melting in the other. The authors suggested that such a scenario was possible during the 40K world because Antarctic glaciers may have melted inland instead of extending to the shoreline and calving into the oceans. Today, ice extends to the coastline all around Antarctica, and the size of the ice sheet is more or less fixed by what the footprint of the continent can bear. If the ice sheet does not reach the ocean, however, it can vary in size from zero to nearly today's magnitude. In Raymo et al.'s scenario (2006), the amplitude of glacial precession cycles would be much smaller in Antarctica than in the Northern Hemisphere.

However, southern glaciers are colder and more depleted in ^{18}O, and hence have a disproportionate influence on the $\delta^{18}O$ of seawater. At the frequency of precession, north and south thus have similar but opposite impacts on $\delta^{18}O$ of seawater (and benthic forams), their effects cancel, and one does not observe precession periodicity in the $\delta^{18}O$ data. On the other hand, northern and southern changes are in phase at the frequency of tilt (warm northern summers coincide with warm southern summers). Their added effects give the record its characteristic 40 Kyr periodicity.

Huybers (2006) proposed a competing hypothesis; he argued that poleward of 60°, the effects of precession are self-cancelling because seasons spent closer to the sun are shorter. When the Northern Hemisphere, for example, is close to the sun, it receives more insolation at summer solstice, which would accelerate the melting of ice sheets. However, summers are shorter in accord with Kepler's second law. One can have short, hot summers or long, cool summers, but total summertime warmth is roughly constant. Thus, Huybers (2006) proposed that precession did not drive growth and decay of Antarctic ice sheets. On the other hand, between 60–45° (the major domain of Northern Hemisphere glaciers during the 100K world), the total amount of heat received during summer is greater when summer solstice is close to the sun. Thus, when Earth cools enough that large North American and Eurasian glaciers extend equatorward of 60° latitude, precession forcing would become more important. This effect may thus explain why precession forcing is more prominent in the $\delta^{18}O$ record of large glacial cycles of the past million years, but absent during the 40K world when glaciers may have been present at higher latitudes.

The Transition to the 100K World, ~1 Ma

At about 1 Ma, the dominant period of glacial-interglacial cycles increased from 40 Kyr to about 100 Kyr. Statistical analysis of the benthic foram $\delta^{18}O$ record shows that variability at the 41 Kyr period of tilt has been important throughout the past 3 Myr. Variability at the frequencies of precession (19 and 23 Ka) and at the 100 Kyr period became important between about 1.4–1.2 Ma. Starting with the interglacial at 1.2 Ma, one can sometimes observe longer climate cycles encompassing 2 or 3 individual tilt cycles (fig. 9.3). Then by 800 Ka, the dominant variance is in climate cycles lasting about 100 Kyr.

Three general paradigms are offered to explain the change from the 40K world to the 100K world. The first is that, around 1 Ma, the Pliocene (or Cenozoic) cooling became intense enough that glaciers could no longer be melted by a single period of high insolation due to tilt. The ultimate cause of the cooling could have been a decrease in CO_2 (Raymo et al. 2006), which might have caused the planet to cool enough to activate another agent in the climate system. For example, lower CO_2 concentrations and modest cooling might have triggered increased sea ice or a cooling of the eastern equatorial Pacific, both of which could have led to more extensive glaciers.

The second paradigm is that the ice ages of the 40K world changed Earth's surface such that deglaciation became more difficult. According to Clark and Pollard (1998) and Clark et al. (2006), this change was the stripping of soil from the Canadian Shield. They envision that the glaciated region of North America started out, around 2.7 Ma, with a soil layer (or "regolith") about 50 m thick. The soft nature of the soil would have served to lubricate the flow of glaciers even though the bottom of the ice sheet would have been frozen. Gradually, however, the flowing ice would entrain and erode the soil. Eventually, the soil would be gone and glaciers would be frozen onto bedrock. They would then have moved much more slowly, they would build to greater heights and volumes, and they would be harder to melt. Upon crossing a threshold where glaciers growing at the start of a 40 Kyr cycle were unable to melt by its end, we had entered the 100K world.

The largest of the glacial ice sheets, the Laurentide, covered the northern latitudes of central and eastern North America (fig. 9.1). There is some evidence that the footprint of this ice sheet during the 40K world was as large as during the 100K world (Clark and Pollard 1998; Clark et al. 2006). Evidence comes from glacial moraines around the southern boundary of the Late Pleistocene ice sheets that have been radiometrically dated to about 2.4 Ma (Balco and Rovey 2010). Also, sediment cores from the Gulf of Mexico contain an isotopic signature of glacial meltwater, originating from ice sheets south of the St. Lawrence River, dating to before 1 Ma. According to benthic foram δ^{18}O data, ice volumes were clearly far greater in the 100K world, and therefore the ice sheets must have been far thicker. As outlined above, the erosion of soil would explain the change.

A third paradigm for the transition from 40 to 100 Kyr periodicity starts with a premise to which we will return shortly: the 100 Kyr cycles are really cycles paced by tilt, with deglaciations occurring every second or third tilt cycle (i.e., after 80 Kyr or 120 Kyr). The paradigm is that the shift in period is a consequence of chaotic variability in the climate system (Huybers 2009). Chaos is at least partially a response to rapid melting of ice during a termination, as well as the fact that the melting rate reflects recent as well as current conditions. At least in a model, the chaotic response to obliquity variations can cause the duration of a single glacial cycle to vary between 1 and 3 tilt cycles (resulting in periods of 40, 80, and 120 Kyr). With the auspicious choice of parameters, a very simple model of the response of the ice sheet to forcing simulates a long string of 40 Kyr cycles followed by a string of 80 or 120 Kyr cycles.

THE NATURE AND DYNAMICS OF THE 100 KYR WORLD, 0.9 MA TO PRESENT

Sawtooth climate cycles recorded by δ^{18}O of Foraminifera

We start this discussion, as for the earlier topics, with a focus on the δ^{18}O of Foraminifera. The first measurements of foram $\delta^{18}O$ were made on Caribbean planktonic forams and presented in the classic papers of Cesare Emiliani (Emiliani 1955, 1958, 1964). Broecker and Van Donk (1970) summarized this work, adding a detailed record for another Caribbean core. They emphasized the assymetric nature of the 100 Kyr cycles, with long intervals of increasing $\delta^{18}O$ (indicating cooling and ice sheet growth), followed by short intervals of decreasing $\delta^{18}O$ (indicating warming and deglaciation). Following this early work, a great deal of data have been collected. Much of our data and modern understanding of foram $\delta^{18}O$ variations comes from the distinguished late Cambridge scientist Sir Nicolas Shackleton. His work, along with many other studies, has confirmed the sawtooth nature of the late Pleistocene climate record. This pattern is clearly evident in the benthic foram curve of Lisiecki and Raymo (2005) (fig. 9.3). It is understood to reflect the fact that ice sheets can melt a lot faster than they can grow (growth requires extensive snow accumulation in a cold and dry climate).

There are some interesting differences between the same phases of different glacial cycles (fig. 9.4). For example, Interglacial Stage 11 (which is incompletely represented in this figure) at ~400 Ka, was much longer than the other interglacials, and the present interglacial has maintained peak temperatures longer than most. Interglacial Stage 7 (245 Ka) appears to have been short and relatively cool. The last interglacial (at about 125 Ka) was considerably warmer than the present interglacial (fig. 9.5), and sea level was about 7 m higher due to a further

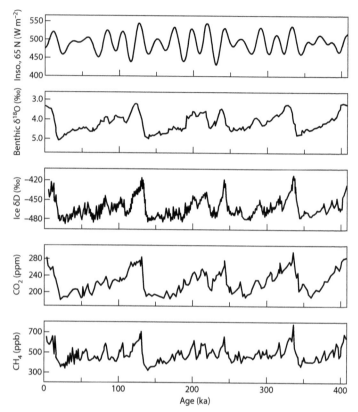

Fig. 9.4. Ice core and related records of climate change, 410 Ka to the present. *Top graph*, average June insolation at 65° N; *next*, the $\delta^{18}O$ of benthic Foraminifera according to the stack of Lisiecki and Raymo (2005); *next*, hydrogen isotopic temperature of ice, (trending warmer toward top of graph); *next*, CO_2; and *bottom*, CH_4; all plotted versus time. Lower benthic $\delta^{18}O$ values (upward on graph) imply warmer deep waters and less ice; and heavier δD values (upward on the graph) imply warmer East Antarctic temperatures.

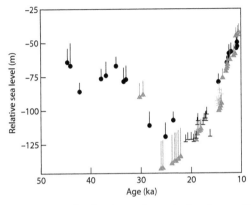

Fig. 9.5. History of sea level and ice volume during the past 50 Ka. The inverted circular and triangular lollipops give the depth below current sea level of fossil corals as a function of age. The solid parts of each symbol give collection depth with respect to modern sea level (corrected for tectonic uplift); the handles go from these depths to the maximum possible contemporaneous sea level, thereby allowing for the fact that corals can live below the sea surface. From the summary in Clark et al. (2009).

retreat of the ice sheets on Greenland and West Antarctica (Kopp et al. 2009).

Globally significant records of climate change over the past 400 Kyr are summarized in figure 9.4 (Blunier et al. 2012). The record of benthic foram $\delta^{18}O$ in this plot represents the latest part of the record in figure. 9.3. Over the past 420 Ka there have been four complete 100 Kyr cycles, with climate shifting from interglacial to glacial conditions and back.

Greenhouse gas concentration changes during the ice ages

Greenhouse gas concentration changes are closely aligned with changes in global glacial conditions as recorded by $\delta^{18}O$ of benthic Foraminifera. Carbon dioxide concentrations are low during glacials, high during interglacials, and intermediate during periods of intermediate climate. They correlate with $\delta^{18}O$ of benthic Foraminifera during excursions lasting tens of thousands of years; for example, the foram $\delta^{18}O$ minima at 290, 218, 205, and 82 Ka are all accompanied by maxima in CO_2. Concentrations of CH_4, an important secondary greenhouse gas, also track the 100 Kyr and 20–40 Kyr variability in temperature even more faithfully than CO_2. Methane (CH_4) is produced by fermentation. Prokaryotes, which lack oxidants, make energy by splitting organic matter into CO_2 and CH_4. Under natural conditions, CH_4 is mostly produced in continental bogs and swamps, where O_2 cannot penetrate through waterlogged soils. These conditions are more prominent in warm times than cold, explaining why CH_4 concentrations are higher in interglacial times. High frequency "noise" in the CH_4 record reflects climate changes over periods of order 1000 years, a topic for the next chapter.

Regional temperature histories recorded in the isotopic composition of ice from ice cores

Just as the oxygen isotope composition of Foraminifera is an important indicator of climate in deep-sea sediment records, the oxygen and hydrogen isotope compositions of ice are important climate indicators in ice core records. Their proxy quality comes from the fact that both O and H isotopes are fractionated when water or ice are in equilibrium with water vapor. In this equilibrium, the heavy isotopes enter preferentially into the condensed phase (i.e., water or ice). As an airmass transits from warm to cold regions, it progressively loses water (or ice) because the saturation vapor pressure decreases with temperature. At each point along the transit, the condensed phase is enriched in the heavy isotope with respect to water gas. As a result, water remaining in the airmass becomes more and more depleted in the heavy isotopes of H and O. The magnitude of the depletion is a measure of the cooling experienced by the airmass at the location of precipitation.

For the Vostok ice core (see fig. 9.4) raised in East Antarctica by a Russian and French team, the best record of ice isotopes comes from hydrogen studies rather than oxygen studies. The relevant heavy isotope is ^{2}H, hydrogen with a proton and a neutron in the nucleus. For historical reasons, this isotope is called deuterium (written "D"). By convention, ^{1}H is simply written as "H." The ^{2}H/^{1}H ratio is δD (see box 1), defined by analogy with δ^{18}O as:

$$\delta D = \{(D/H)_{sample}/(D/H)_{reference} - 1\}*1000$$

The δD of Vostok ice also covaries closely with the δ^{18}O of Foraminifera (fig. 9.4). Lower values of δD in ice signify colder temperatures over East Antarctica. The similarity between the ice δD curve and the foram $\delta^{18}O$ curve indicates that Antarctic climates were cold when the high-latitude oceans were cold and when the Northern Hemisphere continents were heavily glaciated. These results thus show that climate changes occurring over tens of thousands of years were globally synchronous between the hemispheres.

Finally, the dust concentration (increasing downward in fig. 9.4) also covaries with climate. The dust flux to East Antarctica depends on two factors: aridity in the source areas (Southern Hemisphere continental areas, particularly Patagonia), and winds to transport the dust to Antarctica. Modeling studies show that the first is more important than the second, so that high dust accumulation in East Antarctica signifies extensive aridity around Patagonia during glacial times. This condition reflects the fact that glacials were drier than interglacials in most continental areas.

The general pattern of ice age climate change suggests that Earth's climate generally drifts toward the glacial condition. Upon achieving it, however, some threshold is crossed and conditions become unstable, leading to deglaciation (Paillard 1998; Raymo 1997). In the next section, we describe the process of deglaciation, and then examine the climate forcing that led the to 100 Kyr cycles.

Ice age terminations

The greatest extent of the ice sheets during the last glacial maximum, based on both the distribution of glacial deposits and direct evidence for sea level lowering, lasted from approximately 26 to 19 Ka. Figure 9.5 summarizes critical evidence. Of particular importance is the coral record of sea level, from the landmark work of by Bard et al. (1996) and Fairbanks (1989). Much effort was involved recovering fossil corals by drilling offshore of Barbados and other tropical islands, and then precisely dating the corals by uranium series methods. For a given coral age, sea level is approximately equal to the depth from which the coral was collected. It is necessary to make corrections for the fact that corals can live in waters below sea level, and for tectonic uplift of islands like Barbados. The result is a detailed history of sea level and ice volume going back tens of thousands of years (fig. 9.5).

Maximum glaciation and its termination was a complex phenomenon that occurred at different times in different regions Clark et al. (2009). Nevertheless, all the major continental ice sheets reached their greatest extent around 25 Ka, and most were retreating by 22–20 Ka. Lowest sea level coincided with maximum ice extent, as it must. There is a close connection between sea level and solar insolation. June insolation at 65° N was a minimum at 23 Ka, when ice sheets were at or very close to their greatest extent. Melting began as insolation started to rise, was about halfway complete by the time insolation reached its maximum (~11.5 Ka), and continued into the early Holocene as summertime insolation slowly decreased. At the glacial maximum, sea level was about 125 m below the modern level; it rose to within about 10 m of the modern level by 8 Ka. The lag in sea level with respect to insolation reflects the response time of the ice sheets to changing insolation, greenhouse forcing, and other influences on climate. Sea level, as deduced from the $\delta^{18}O$ of benthic Foraminifera, agrees closely with sea level as inferred from corals (Peltier and Fairbanks 2006; Waelbroeck et al. 2002).

Three important factors drove the transition from a glacial to interglacial climate. First, rising summertime insolation at 65° N promoted melting of the ice sheets. Second, melting of the ice sheets lowered albedo, introducing a positive feedback in which more warming caused more melting, leading to more warming, and so forth. Third, concentrations of CO_2 and CH_4 rose, promoting additional warming. Thus a series of positive feedback loops sustained deglaciation until large ice sheets remained only on Greenland and Antarctica. At that point, a stable climate was reached that has now lasted about 7 Kyr.

The warming with deglaciation was itself a complex event that was punctuated by acceleration in some domains and stasis or even climate reversals in others. According to the isotopic composition of ice cores, some polar regions began warming well after sea level began to rise. East Antarctica began warming at about 17.5 Ka, and Greenland about 3 Kyr later (see curves A and C in fig. 9.2). The dramatic Greenland warming at 14.6 Ka was followed by the climate deterioration of the Younger Dryas, a cold period from 12.8 to 11.6 Ka. The Younger Dryas was followed by another episode of rapid warming.

Meanwhile, warming with deglaciation in Antarctica was, to some extent, opposite to that of Greenland; East Antarctica warmed precisely during those intervals when Greenland was cold. These episodes represent deglaciation punctuated by rapid climate change events, in which sudden variations in ocean circulation influenced regional climates in different ways (as discussed in the following chapter). One manifestation of these climate changes is the variation in the $\delta^{18}O$ of $CaCO_3$ of speleothems from East Asian caves (Hulu and Donge records, plot D in fig. 9.2; see also box 3). These observations reflect changes in East Asian hydrology associated with high latitude climate oscillations discussed above.

It is interesting to compare the sequence of climate changes during the last glacial termination with those of the second, third, and fourth glacial terminations, going back in time. Based on the benthic foram $\delta^{18}O$ record, at each termination, sea level began to rise, and the high latitudes and deep oceans began to warm, within a few thousand years of the minimum in insolation. The CO_2 rise with deglaciation, and the warming of East Antarctica , closely track the $\delta^{18}O$ decrease. Dust accumulation in Vostok decreased from glacial to interglacial time as temperate regions of the Southern Hemisphere became wetter. Higher CH_4 concentrations indicate warmer and wetter climates in the Northern Hemisphere. Methane rose during the early stages of deglaciation and jumped when CO_2 concentrations and East Antarctic temperatures reached maxima. Based on results from the past 100 Kyr (see next chapter), the CH_4 jump accompanied warming in the North Atlantic and surrounding regions. During terminations, this warming generally put an end to the discharge of tidewater glaciers and icebergs into the North Atlantic as marked by the sharp decrease of rock grains carried by icebergs ("ice-rafted detritus"). Finally, the rise in CO_2 with deglaciation closely tracked Antarctic warming.

While glacial terminations differ in detail, they follow a basic sequence of events. Glacial terminations begin when summertime boreal insolation begins rising. Melting of ice sheets and warming of the deep ocean are roughly synchronous with the warming of Antarctica and the rise of atmospheric CO_2. Northern Hemisphere climates warm and become wetter toward the end of a glacial termination, when CO_2 has reached its maximum value. The link between Antarctic temperature, deep ocean temperature/ice volume, and CO_2 suggests that the Southern Ocean plays a large role in the CO_2 rise during deglaciation. The late and rapid warming of Greenland, and of the Northern Hemisphere more generally, indicate that the ice sheets themselves can maintain cold climates in the north until they shrink below a threshold value. The connection between insolation change and deglaciation strongly supports a big role for Milankovitch (orbital) forcing in the pacing of glacial-interglacial climate change (CLIMAP 1976).

Interglacial climates are impermanent. Eventually, orbital changes will induce cooling, and the opposite feedback loops will push the planet toward a glacial mode. Whether Earth is heading toward warmer or cooler climates, the feedbacks run out of steam at some point, limiting the extremes to something like today's world on the warm end, or at the other end, the world of the last glacial maximum described above.

ORBITALLY DRIVEN CLIMATE CHANGE INDEPENDENT OF THE ICE SHEETS

Superimposed on the 100 Kyr cycles are two other modes of climate change. The first consists of events lasting hundreds or thousands of years that are global in extent but are most strongly manifested in the North Atlantic region. These events involve very rapid climate change, are prominent during glacial times, but may also have muted expressions during interglacials. They are the subject of the next chapter (chapter 10); interglacial rapid change events are discussed briefly in the chapter on the Holocene (chapter 11).

The second climate mode is paced primarily by precession and occurs mainly in the lower latitudes. The most dramatic example of this mode is in hydrologic changes associated with the East Asian monsoon. Variations in the $\delta^{18}O$ of $CaCO_3$ making up south China speleothems (cave deposits) vary strongly with precession and with Northern Hemisphere summer insolation (Y. J. Wang 2008; fig. 9.6). These $\delta^{18}O$ changes originate from changes in isotope fractionation associated with the evaporation and condensation that lead to the rainfall from which $CaCO_3$ precipitates. Speleothem $\delta^{18}O$ changes thus record regional variations in amounts,

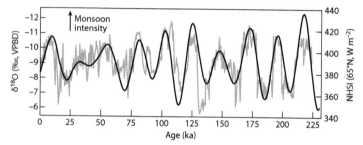

Fig. 9.6. Black curve: July 21 insolation at 65° N. Gray curve: $\delta^{18}O$ of $CaCO_3$ in two East Asian caves. Isotopically lighter (more negative) values correspond to more summer monsoon rainfall.
From Y. J. Wang (2008).

sources, and pathways of precipitation (e.g., Dayem et al. 2010). Most workers attribute $\delta^{18}O$ variations to changes in the amount of summertime precipitation associated with the East Asian monsoon; low $\delta^{18}O$ values signify more intense monsoon and more rainfall (e.g., Y. J. Wang 2008). We follow this interpretation but note that it was challenged by Dayem et al. (2010), who argued that other aspects of the hydrologic cycle could have a major influence on the relation between precipitation and $\delta^{18}O$.

The tight correlation between $\delta^{18}O$ in Chinese speleothems and Northern Hemisphere summer insolation, or precession (fig. 9.6), thus suggests that the summer monsoon in East Asia was stronger when summertime insolation was more intense (Y. J. Wang 2008). This connection has two causes. First, more insolation means hotter continental temperatures during summertime, leading to stronger monsoons. Second, more insolation means stronger steering of the Intertropical Convergence Zone (ITCZ) to the north during boreal summer. The ITCZ is the band in the tropics where equatorward-flowing air in both hemispheres collides and rises, leading to high rainfall. In the present context, the northward shift of the ITCZ leads to greater summer precipitation.

At low latitudes, precession introduces more variability than tilt to summertime insolation. Since insolation variations due to precession are out of phase between the hemispheres, we might conjecture that monsoonal precipitation changes in the Southern Hemisphere will be out of phase with respect to changes in the Northern Hemisphere. There is evidence to support this view. For example, X. F. Wang et al. (2006) studied a cave in southeastern Brazil that falls in the domain of the South American monsoon. Over the past 36 Ka, $\delta^{18}O$ of speleothems from this cave closely follows local summer insolation, which changed mainly due to precession. However, like precession warming, $\delta^{18}O$ is out of phase between Northern and Southern Hemispheres. In both China and Brazil, the speleothems are isotopically light when local summer insolation is highest, signifying a stronger monsoon and increased precipitation at these times.

Another example of orbitally paced changes, discussed in the next chapter, is precession-paced variations of temperature and precipitation changes observed in the Holocene.

WHAT DRIVES THE 100 KYR ICE AGE CYCLES?

Our emerging understanding of the dynamics of the 100 Kyr climate cycle rests on four concepts. First, ice sheets respond to Milankovitch climate forcing. When summers are warm, ice sheets tend to retreat, and when summers are cool, ice sheets tend to advance. Second, Earth's favored climate mode is toward the colder end of the spectrum. The main evidence for this comes from the benthic foram $\delta^{18}O$ curves, showing that Earth has most often been in the cold mode while trending toward glacial maximum conditions. Third, ice volume has a high-volume threshold that, when crossed, leads to instability and rapid deglaciation (Paillard 1998; Raymo 1997). Fourth, changes in the physical climate system cause the atmospheric CO_2 concentration to rise when Earth is warming and fall when Earth is cooling. This CO_2 response provides a positive feedback that leads to the full observed amplitude of glacial-interglacial cycles.

Consider the curve of benthic foram $\delta^{18}O$ versus age in figure 9.4. As explained in box 2, increasing $\delta^{18}O$ signifies cooler temperatures and larger continental ice sheets. As insolation fell from 129 through 116 Ka, there was a significant trend toward glacial climates. By ~112 Ka, $\delta^{18}O$ reached half of that at glacial maximum. The $\delta^{18}O$ then oscillated, anticorrelated with Northern Hemisphere June insolation. June insolation was at a maximum at 106 Ka and 84 Ka; these times corresponded to $\delta^{18}O$ minima. The $\delta^{18}O$ increased from 83 to 54 Ka, flattened, and began increasing again as June insolation started falling at 34

Ka. The $\delta^{18}O$ then began to diminish rapidly at ~20 Ka, as insolation increased. We do not see cooling and growth of ice corresponding to the insolation decrease during the Holocene. Nevertheless, the record clearly establishes the close link between ice volume or sea level change, and summer insolation. This link is also supported by climate models (Huybers and Tziperman 2008; Jackson and Broccoli 2003), although these models may not simulate the full magnitude of the glacial cooling.

The benthic $\delta^{18}O$ record is also consistent with the idea that complete deglaciation of temperate and subpolar continents requires the crossing of a threshold in ice volume. For example, summer insolation was higher at 106 Ka and 84 Ka than at 21 Ka. However, the earlier insolation maxima do not lead to complete deglaciation. That happens only once the ice sheets have reached their critical extent. The situation was similar for previous glacial cycles. The response of ice volume to summertime insolation is intuitively reasonable. The main challenge, then, is to understand the factors leading CO_2 to change (approximately) in phase with global climate, and to understand the nature of the ice volume threshold for deglaciation.

Near the beginning of this chapter, we discussed glacial-interglacial fluctuations in the CO_2 concentration of air. In glacial times, slower overturning of the Southern Ocean, greater windblown dust supply of fertilizing iron, colder ocean temperatures, and deep-sea $CaCO_3$ compensation all tend to lower pCO_2. Higher ocean salinity and the retreat of the land biosphere during ice ages tend to increase pCO_2. The processes lowering pCO_2

win, pCO_2 is lower during the ice ages, and there is a positive feedback involving increased ice volume, increased albedo, lower atmospheric CO_2, and colder climates.

Arguably, the key event linking CO_2 to climate is the one that initiates changes in the rate at which waters of the ocean interior come to the surface in the Southern Ocean, thereby releasing to the atmosphere CO_2 that accumulated from the decay of sinking organic matter. This rate seems to be affected by the status of deep water formation in the North Atlantic. During terminations, there may be intervals in which North Atlantic Deep Water does not form, because meltwater runoff lowers the salinity. At these times, Southern Ocean mixing is enhanced; one possible mechanism is wind belts shifting south with westerly winds becoming stronger over the Southern Ocean (Anderson et al. 2009; Cheng et al. 2009; Sigman et al. 2010; Toggweiler and Lea 2010). The higher flux of CO_2-rich waters to the surface leads to the rise in atmospheric CO_2 that promotes deglaciation.

The remaining question concerns the nature of the ice volume threshold for deglaciation. There seem to be two candidates influencing this threshold. First, growth of an ice sheet depresses the underlying basement rocks ("isostatic adjustment"), because the weight causes the underlying crust to flow laterally away from the ice. The margins of the ice sheet then tend to sink with respect to sea level, making it easier for the oceans to erode glaciers on the borders of the continents. Second, the bed of the glacier paradoxically warms as the thickness of ice increases. The reason is that the ice sheet has the effect of insulating the bed from cold air at the surface of the glacier. When the ice is thick, geothermal heating can warm the bed and melt basal ice, greatly facilitating destruction of the glacier. Modeling studies have demonstrated the plausibility of this mechanism (Marshall and Clark 2002).

From this discussion we can create a simple, provisional model of the 100 Kyr cycle. An interglacial period will end, and ice sheets will start to grow, when a minimum in Northern Hemisphere summer insolation is sufficiently intense. Ice sheets will wax and wane in concert with summer insolation changes, but tend to get larger over time. At the same time, the atmospheric CO_2 concentration will fall as progressive Northern Hemisphere cooling causes increased stratification and slower upwelling in the Southern Ocean. Lower CO_2 further cools the planet and provides a positive feedback to the growth of ice sheets. Eventually ice volume crosses a threshold leading to instability. The next strong maximum in Northern Hemisphere summer insolation will lead to the melting of ice sheets and warming of the planet (Raymo, 1977). Events associated with deglaciation can cause increased mixing of deep waters back to the surface of the Southern Ocean, leading to increased atmospheric CO_2 burdens, rapid deglaciation, and the return to interglacial conditions.

REFERENCES

Papers with asterisks are suggested for further reading.

Anderson, R. F., S. Ali, L. I. Bradtmiller, S. H. H. Nielsen, M. Q. Fleisher, B. E. Anderson, and L. H. Burckle (2009),

Wind-driven upwelling in the Southern Ocean and the deglacial rise in atmospheric CO_2, *Science*, *323*, 1443–1448.

Balco, G., and C. W. Rovey, II (2010), Absolute chronology for major Pleistocene advances of the Laurentide Ice Sheet, *Geology*, *38*(9), 795–798.

Ballantyne, C. K. (2010), Extent and deglacial chronology of the last British-Irish Ice Sheet: Implications of exposure dating using cosmogenic isotopes, *Journal of Quaternary Science*, *25*(4), 515–534.

Blunier, T., M. L. Bender, B. Barnett, and J. C. von Fischer (2012), Planetary fertility during the past 400 ka based on the triple isotope composition of O_2 in trapped gases from the Vostok ice core, *Climate of the Past*, *9*, 1509–1526.

Braconnot, P., B. Otto-Bliesner, S. Harrison, S. Joussaume, J.-Y. Petterschmitt, A. Abe-Ouchi, M. Crucifix, et al. (2007), Results of PMIP2 coupled simulations of the Mid-Holocene and Last Glacial Maximum—Part 1: Experiments and large-scale features, *Climate of the Past*, *3*, 261–277.

Broecker, W. S. (1982), Ocean chemistry during glacial time, *Geochimica et Cosmochimica Acta*, *46*, 1689–1705.

Broecker, W. S., and P. C. Orr (1958), Radiocarbon Chronology of Lake Lahontan and Lake Bonneville, *Geological Society of America Bulletin*, *69*, 1009–1032.

Broecker, W. S., and A. Kaufman (1965), Radiocarbon chronology of Lake Lahontan and Lake Bonneville II, Great Basin, *Geological Society of America Bulletin*, *76*, 537–566.

Broecker, W. S., and J. Van Donk (1970), Insolation changes, ice volumes, and O-8 record in deep-sea cores, *Reviews of Geophysics and Space Physics*, *8*, 169–198.

Broecker, W. S., G. H. Denton, M. E. Edwards, H. Cheng, R. B. Alley, and A. E. Putnam (2010), Putting the Younger Dryas cold event into context, *Quaternary Science Reviews*, *29*, 1078–1081.

Cane, M. A., and P. Molnar (2001), Closing of the Indonesian seaway as a precursor to east African aridification around 3–4 million years ago, *Nature*, *411*, 157–162.

Cheng, H., R. L. Edwards, W. S. Broecker, G. H. Denton, X. Kong., Y. Wang, R. Zhang, and X. Wang (2009), Ice Age terminations, *Science*, *326*, 248–252.

*Clark, P. U., and D. Pollard (1998), Origin of the middle Pleistocene transition by ice sheet erosion of regolith, *Paleoceanography*, *13*(1), 1–9.

Clark, P. U., D. Archer, D. Pollard, J. D. Blum, J. A. Rial, V. Brovkin, A. C. Mix, et al. (2006), The middle Pleistocene transition: Characteristics, mechanisms, and implications for long-term changes in atmospheric pCO_2, *Quaternary Science Reviews*, *25*, 3150–3184.

Clark, P. U., A. S. Dyke, J. D. Shakun, A. E. Carlson, J. Clark, B. Wohlfarth, J. X. Mitrovica, et al. (2009), The Last Glacial Maximum, *Science*, *325*, 710–714.

*CLIMAP Project Members (1976), The surface of the ice-age Earth, *Science*, *191*, 1131–1137.

Colinveaux, P. A., P. E. De Oliveira, and M. B. Bush (2000), Amazonian and neotropical plant communities on glacial time-scales: The failure of the aridity and refuge hypotheses, *Quaternary Science Reviews*, *19*, 141–169.

Conway, H., B. L. Hall, G. H. Denton, A. M. Gades, and E. D. Waddington (1999), Past and future grounding-line retreat of the West Antarctic ice sheet, *Science*, *286*, 280–283.

Dayem, K. E., P. Molnar, D. S. Battisti, and G. H. Roe (2010), Lessons learned from oxygen isootpes in modern precipitation applied to interpretation of speleothem records of paleoclimate from eastern Asia, *Earth and Planetary Science Letters*, *295*, 219–230.

de Vernal, A., F. Aynaud, A. Henry, C. Hillaire-Marcel, L. Londeix, S. Mangin, J. Mattheissen, et al. (2005), Reconstruction of sea-surface conditions at middle to high latitudes of the Northern Hemisphere during the Last Glacial Maximum (LGM) based on dinoflagellate cyst assemblages, *Quaternary Science Reviews*, *24*, 897–924.

Delmas, R. J., J.-M. Ascenio, and M. Legrand (1980), Polar ice evidence that atmospheric CO_2 20,000 yr BP was 50% of present, *Nature*, *284*, 155–157.

deMenocal, P. B. (2004), African climate change and faunal evolution during the Pliocene-Pleistocene, *Earth and Planetary Science Letters*, *220*, 3–24.

Dowsett, H. J., M. M. Robinson, D. K. Stoll, and K. M. Foley (2010), Mid-Piacenzian mean annual sea surface temperature analysis for data-model comparisons, *Stratigraphy*, *7*, 189–198.

Dowsett, H. J., M. M. Robinson, A. Haywood, U. Salzmann, D. Hill, L. Sohl, M. Chandler, et al. (2010), The PRISM3D, *Stratigraphy*, *7*, 123–139.

Dubois, N., M. Kienast, C. Normandeau, and T. D. Herbert (2009), Eastern equatorial Pacific cold tongue during the Last Glacial Maximum as seen from alkenone paleothermometry, *Paleoceanography*, *24*(PA4207), 1–12.

Dyke, A. S. (1999), Lat Glacial Maximum and deglaciation of Devon Island, Arctic Canada: Support for an Innuitian Ice Sheet, *Quaternary Science Reviews*, *18*, 393–420.

Emiliani, C. (1955), Pleistocene temperatures, *Journal of Geology*, *63*, 538–578.

Emiliani, C. (1958), Paleotemperature analysis of core-280 and Pleistocene correlations, *Journal of Geology*, *66*, 264.

Emiliani, C. (1964), Paleotemperature analysis of the Caribbean cores A-254-BR-C and CP-28, *Bulletin of the Geological Society of America*, *75*, 129–143.

Etourneau, J., R. Schneider, T. Blanz, and P. Martinez (2010), Intensification of the Walker and Hadley atmospheric circulations during the Pliocene-Pleistocene climate transition, *Earth and Planetary Science Letters*, *297*, 103–110.

*Fedorov, A. V., P. S. Dekens, M. McCarthy, P. B. deMenocal, M. Barreiro, R. C. Pacanowski, and S. G. Philander (2006), The Pliocene paradox (mechanisms for a Permanent El Niño), *Science*, *312*, 1485–1489.

Gersonde, R., X. Crosta, A. Abelmann, and L. Armand (2005), Sea-surface temperature and sea ice distribution of the Southern Ocean at the EPILOG Last Glacial Maximum: A circum-Antarctic view based on siliceous microfossil records, *Quaternary Science Reviews*, *24*, 869–896.

Hall, B. L., G. H. Denton, and C. H. Hendy (2000), Evidence from Taylor Valley for a grounded ice sheet in the Ross Sea, Antarctica, *Geografiska Annaler*, *82A*, 275303.

Harrison, S. P., and I. C. Prentice (2003), Climate and CO_2 controls on global vegetation distribution at the last glacial maximum: Analysis based on palaeovegetation data, biome modelling and palaeoclimate simulations, *Global Change Biology*, *9*, 983–1004.

Haug, G. H., A. Ganapolski, D. Sigman, A. Rosell-Mele, G.E.A. Swann, R. Thiedemann, S. L. Jaccard, et al. (2005), North

Pacific seasonality and the glaciation of North America 2.7 million years ago, *Nature*, *433*, 821–825.

Henrot, A.-J., L. Francois, S. Brewer, and G. Munhoven (2009), Impacts of land surface properties and atmospheric CO_2 on the Last Glacial Maximum climate: A factor separation analysis, *Climate of the Past*, *5*, 183–202.

*Huybers, P. (2006), Early Pleistocene glacial cycles and the integrated summer insolation forcing, *Science*, *313*, 508–511.

Huybers, P. (2009), Pleistocene glacial variability as a chaotic response to obliquity forcing, *Climate of the Past*, *5*, 481–488.

Huybers, P., and P. Molnar (2007), Tropical cooling and the onset of North American glaciation, *Climate of the Past*, *3*, 549–557.

Huybers, P., and E. Tziperman (2008), Integrated summer insolation forcing and 40,000-year glacial cycles: The perspective from an ice-sheet/energy balance model, *Paleoceanography*, *23*. doi: 10.1029/2007PA001463.

Jackson, C. S., and A. J. Broccoli (2003), Orbital forcing of Arctic climate: Mechanisms of climate response and implications for continental glaciation, *Climate Dynamics*, *21*, 539–557.

Jahn, A., M. Claussen, A. Ganopolski, and V. Brovkin (2005), Quantifying the effect of vegetation dynamics on the climate of the Last Glacial Maximum, *Climate of the Past*, *1*, 1–7.

Jansen, E., T. Fronval, F. Rack, and J.E.T. Channell (2000), Pliocene-Pleistocene ice rafting history and cyclicity in the Nordic Seas during the last 3.5 Myr, *Paleoceanography*, *15*(6), 709–721.

Jost, A., D. J. Lunt, M. Kageyama, A. Abe-Ouchi, O. Peyron, P. J. Valdes, and G. Ramstein (2005), High-resolution simulations of the last glacial maximum climate over Europe:

A solution to discrepancies with continental palaeoclimatic reconstructions, *Climate Dynamics*, *24*, 577–590.

Karas, C., D. Nurnberg, A. K. Gupta, R. Tiedemann, K. Mohan, and T. Bickert (2009), Mid-Pliocene climate change amplified by a switch in Indonesian subsurface throughflow, *Nature Geoscience*, *2*, 434–438.

Kleiven, H. F., E. Jansen, T. Fronval, and T. M. Smith (2002), Intensification of Northern Hemisphere glaciations in the circum Atlantic region (3.5–2.4 Ma)—ice-rafted detritus evidence, *Palaeogeography, Palaeoclimatology, Palaeoecology*, *184*, 213–223.

Kohler, P., H. Fischer, G. Munhoven, and R. E. Zeebe (2005), Quantitative interpretation of atmospheric carbon records over the last glacial termination, *Global Biogeochemical Cycles*, *19*(GB4020), 1–24.

Kopp, R. E., F. J. Simons, J. X. MItrovica, A. C. Maloof, and M. Oppenheimer (2009), Probabilistic assessment of sea level during the last interglacial stage, *Nature*, *462*, 863–867.

Lisiecki, L. E., and M. E. Raymo (2005), A Pliocene-Pleistocene stack of 57 globally distributed benthic delta O-18 records, *Paleoceanography*, *20*. doi: 10.1029/2004PA001071.

Luz, B., and E. Barkan (2011), The isotopic composition of atmospheric oxygen, *Global Biogeochemical Cycles*, *25*. doi: 10.1029/2010GB003883.

Marchant, R., A. Cleef, S. P. Harrison, H. Hooghiemstra, V. Markgraf, J. van Boxel, T. Ager, et al. (2009), Pollen-based biome reconstructions for Latin America at 0, 6000 and 18 000 radiocarbon years ago, *Climate of the Past*, *5*, 725–767.

MARGO (2009), Constraints on the magnitude and patterns of ocean cooling at the Last Glacial Maximum, *Nature Geoscience*, *2*, 127–132.

Mark, B. G., S. P. Harrison, A. Spessa, M. New, D. J. A. Evans, and K. F. Helmens (2005), Tropical snowline changes at the last glacial maximum: A global assessment, *Quaternary International 138-139*, 168–201.

*Marshall, S. J., and P. U. Clark (2002), Basal temperature evolution of North American ice sheets and implications for the 100-kyr cycle, *Geophysical Research Letters*, *29*. doi:10.1029/2002GL015192.

*Martin, J. H. (1990), Glacial-interglacial CO_2 change: The iron hypothesis, *Paleoceanography*, *5*(1), 1–13.

Martinez-Garcia, A., A. Rosell-Mele, E. L. McClymont, R. Gersonde, and G. Haug (2010), Subpolar link to the emergence of the modern equatorial Pacific cold tongue, *Science*, *328*, 1550–1553.

Maslin, M. A., X. S. Li, M.-F. Loutre, and A. Berger (1998), The contribution of orbital forcing to the progressive intensification of Northern Hemisphere glaciation, *Quaternary Science Reviews*, *17*, 411–426.

Molnar, P. (2008), Closing of the Central American seaway and the ice age: A critical review, *Paleoceanography*, *23*(PA2201), 1–15.

Monnin, E., A. Indermuhle, A. Dallenbach, J. Fluckiger, B. Stauffer, T. F. Stocker, D. Raynaud, and J.-M. Barnola (2001), Atmospheric CO_2 concentrations over the last glacial termination, *Science*, *291*, 112–114.

Mudelsee, M., and M. E. Raymo (2005), Slow dynamics of the Northern Hemisphere glaciation, *Paleoceanography*, *20*(PA4022), 1–14.

Neftel, A., H. Oeschger, J. Schwander, B. Stauffer, and R. Zunbrunn (1982), Ice core sample measurements give atmospheric CO_2 content during the past 40,000 years, *Nature*, *295*, 220–223.

Pagani, M., Z. Liu, J. LaRiviere, and A. C. Ravelo (2009), High Earth-system climate sensitivity determined from Pliocene carbon dioxide concentrations, *Nature Geoscience Advance Online Publication*, 1–7. doi: 10.1038/NGEO724.

*Paillard, D. (1998), The timing of Pleistocene glaciations from a simple multiple-state climate model, *Nature*, *391*, 378–381.

Partridge, T. C., P. B. deMenocal, S. A. Lorentz, M. J. Paiker, and J. C. Vogel (1997), Orbital forcing of climate over South Africa: A 200,000-year rainfall record from the Pretoria saltpan, *Quaternary Science Reviews*, *16*, 1125–1133.

Peltier, W. R. (2004), Global glacial isostasy and the surface of the ice-age earth: The ICE-5G (VM2) model and GRACE, *Annual Review of Earth and Planetary Sciences*, *32*, 111–149.

Peltier, W. R., and R. G. Fairbanks (2006), Global glacial ice volume and Last Glacial Maximum duration from an extended Barbados sea level record, *Quaternary Science Reviews*, *25*, 3322–3337.

Pflaumann, U., M. Sarnthein, M. Chapman, L. D'Abreau, B. Funnell, M. Huels, T. Kiefer, et al. (2003), Glacial North Atlantic: Sea-surface conditions reconstructed by GLAMAP 2000, *Paleoceanography*, *18*(3), 1–20.

Polyak, L., F. Niessen, V. Gataullin, and V. Gainanov (2008), The eastern extent of the Barents-Kara ice sheet during the Last Glacial Maximum based on seismic-reflection data from the eastern Kara Sea, *Polar Research*, *27*, 162–174.

Prentice, I. C., D. Jolly, and BIOME 6000 participants (2000), Mid-Holocene and glacial-maximum vegetation geography of the northern continents and Africa, *Journal of Biogeography*, *27*, 507–519.

Ravelo, A. C. (2010), Warmth and glaciation, *Nature Geoscience*, *3*, 672–674.

Raymo, M. E. (1997), The timing of major climate terminations, *Paleoceanography*, *12*, 577–585.

Raymo, M. E., L. E. Lisiecki, and K. H. Nisancioglu (2006), Plio-Pleistocene ice volume Antarctic climate, and the global $\delta^{18}O$ record, *Science*, *313*, 492–495.

Rojas, M., P. Moreno, M. Kageyama, M. Crucifix, C. D. Hewitt, A. Abe-Ouchi, R. Ohgaito, et al. (2009), The southern westerlies during the last glacial maximum in PMIP2 simulations, *Climate Dynamics*, *32*, 525–548.

Salzmann, U., A. M. Haywood, D. J. Lunt, P. J. Valdes, and D. J. Hill (2008), A new global biome reconstruction and data-model comparison for the Middle Pliocene, *Global Ecology and Biogeography*, *17*, 432–447.

Sarmiento, J. L., and J. R. Toggweiler (1984), A new model for the role of the oceans in determining atmospheric pCO_2, *Nature*, *308*, 621–624.

Seki, O., G. L. Foster, D. N. Schmidt, A. Mackensen, K. Kawamura, and R. D. Pancost (2010), Alkenone and boron-based Pliocene pCO_2 records, *Earth and Planetary Science Letters*, *292*, 201–211.

Shackleton, N. J., J. Backmann, H. Zimmerman, D. V. Kent, M. A. Hall, D. G. Roberts, D. Schnitker, et al. (1984), Oxygen isotope calibration of the onset of ice-rafting and history of glaciation in the North Atlantic region, *Nature*, *307*, 620–623.

*Sigman, D. M., M. P. Hain, and G. H. Haug (2010), The polar ocean and glacial cycles in atmospheric CO_2 concentration, *Nature*, *466*, 47–55.

St. John, K.E.K., and L. A. Krissek (2002), The late Miocene to Pleistocene ice-rafting history of southeast Greenland, *BOREAS*, *31*, 28–35.

Stute, M., M. Forster, H. Frischkorn, A. Serejo, J. F. Clark, P. Schlosser, W. S. Broecker, and G. Bonani (1995), Cooling of tropical Brazil (5°C) during the Last Glacial Maximum, *Science*, *269*, 379–383.

Tziperman, E., and H. Gildor (2003), On the mid-Pleistocene transition to 100-kyr glacial cycles and the asymmetry between glaciation and deglaciation times, *Paleoceanography*, *18*(1), 1–8.

Vizcaino, M., S. Rupper, and J.C.H. Chiang (2010), Permanent El Niño and the onset of Northern Hemisphere glaciations: Mechanism and comparison with other hypotheses, *Paleoceanography*, *25*(PA2205), 1-20.

Waelbroeck, C., L. Labeyrie, E. Michel, J. C. Duplessy, J. F. McManus, K. Lambeck, E. Balbon, and M. Labracherie (2002), Sea-level and deep water temperature changes derived from benthic foraminifera isotopic records, *Quaternary Science Reviews*, *21*, 295-305.

Wang, X. F., A. S. Auler, R. L. Edwards, H. Cheng, E. Ito, and M. Solheld (2006), Interhemispheric anti-phasing of rainfall during the last glacial period, *Quaternary Science Reviews*, *25*, 3391-3403.

*Wang, Y. J. (2008), Millenial- and orbital-scale changes in the East Asian monsoon over the past 224,000 years, *Nature*, *451*, 1090-1093.

Wara, M. W., A. C. Ravelo, and M. L. Delaney (2005), Permanent El Niño-like conditions during the Pliocene warm period, *Science*, *309*, 758–761.

Wolff, E. W., H. Fischer, F. Fundel, U. Ruth, B. Twarloh, G. C. Littot, R. Mulraney, et al. (2006), Southern Ocean sea-ice extent, productivity and iron flux over the past eight glacial cycles, *Nature*, *440*, 491–496.

Wu, H., J. Guiot, S. Brewer, and Z. Guo (2007), Climatic changes in Eurasia and Africa at the last glacial maximum and mid-Holocene: Reconstruction from pollen data using inverse vegetation modelling, *Climate Dynamics*, *29*, 211–229.

10 RAPID CLIMATE CHANGE DURING THE LAST GLACIAL PERIOD

THE LAST ICE AGE WAS MARKED BY A SERIES OF CLImate cycles that, while most intense in the North Atlantic, had manifestations throughout the world. These events have been broadly characterized, and their dynamics have been studied extensively with climate models. Much is known about the sequence of events associated with the cycles. However, basic causes remain to be understood.

Many aspects of the rapid climate changes are illustrated in figure 10.1, which shows five climate records dating from 10 to 60 Ka. Moving forward in time, this interval begins shortly after the start of the last ice age, when the Northern Hemisphere ice sheets grew to more than half of their maximum extent. It ends after Greenland warmed to its postglacial value. The upper three curves are proxy records of Antarctic temperature, based on the isotopic composition of the ice in the Byrd, EDML (EPICA Dronning Maud Land, Antarctica), and EDC (EPICA Dome C, Antarctica) ice cores. The fourth plot is the isotopic temperature of the NGRIP (northern Greenland) ice core. The bottom curve is the atmospheric methane concentration as registered in Greenland and

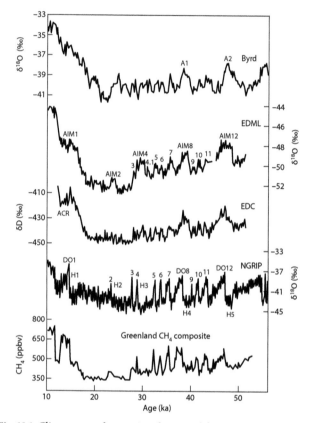

Fig. 10.1. Climate records covering the period from 10 to 60 Ka (Clement and Peterson 2008). In descending order: the $\delta^{18}O$ or δD of ice (both indicating paleotemperature) from the Antarctic ice cores Byrd, EDML (EPICA Dronning Maud Land), and EDC (EPICA Dome C); the $\delta^{18}O$ of the NGRIP (North Greenland) ice core; and the CH_4 concentration as inferred from Greenland ice cores. In the EDML plot, "AIM" indicates Antarctic Isotope Maximum." In the NGRIP plot, unlabeled numbers, or numbers labeled "DO", are the "Interstadial number," and H numbers are the numbers of the respective Heinrich events.

Antarctic ice cores. The interval ends at 10 Ka, when deglaciation was largely complete.

Without any more information, one can make remarkable statements about ice age climates based on this plot. There were more than 12 times when Greenland temperature (represented by NGRIP) warmed *very rapidly*—on the order of decades or less—followed by a slow cooling back to baseline glacial temperatures, followed shortly by another warming. Five intervals, labeled H1–H5 in the NGRIP record, were particularly cold and long. Each warm event in Greenland had a counterpart in the Antarctic record. This connection is clearest when comparing the Antarctic EDC ice core with the Greenland curves. However, Antarctica and Greenland did not precisely track. Instead, Antarctica warmed first. For a time during each event, Greenland remained cold while Antarctica warmed, then Greenland warmed abruptly as both regions attained their maximum temperature. The two regions then cooled more or less synchronously. Figure 10.1 also shows that the atmospheric CH_4 concentration rose each time Greenland warmed abruptly, and remained high for the duration of the Greenland warming.

From these observations we can draw a number of conclusions. First, the rapid climate change events summarized by these data affected polar regions in both hemispheres. Second, rapid climate change events were fast, although more so in the north. Third, continental areas were affected because continental wetlands are the major source of atmospheric CH_4. The close tracking of

CH_4 with Greenland temperature signifies that, in general, wetlands were more extensive and precipitation was greater when Greenland was warm. Finally, climate continually changed rapidly over most of the time between 10–60 Ka (fig. 10.1) and beyond.

We can systematize the information in figure 10.1 by invoking three tightly linked sets of occurrences. Interstadial events are the warm events that repeatedly punctuate the ice core records (numbered DO-1 to DO-12 in the plot of NGRIP $\delta^{18}O$). In extrapolar regions, waters were warmer in the North Atlantic and Indian Oceans during interstadial events, East Asian monsoons were stronger, ocean circulation off the coast of California was less vigorous, and precipitation was greater along the north coast of South America. Interstadial events were called Dansgaard-Oeschger (DO) events by W. S. Broecker in honor of the Swiss and Danish scientists who discovered the evidence for rapid climate change in Greenland ice cores.

The second set of occurrences is the Heinrich events, named again by Broecker after the scientist who first documented their nature. They are labeled H-1 to H-5 on the NGRIP curve in figure 10.1. Heinrich events are periods of about 1–2 Kyr during which Greenland climates were at their coldest levels, sea ice extent dramatically increased in the North Atlantic, and precipitation was elevated in the western United States and southern Brazil. Paradoxically, sea ice increased and sea level rose during Heinrich events as continental ice sheets melted. Massive icebergs were discharged; they carried glacial debris out to the middle of the Atlantic Ocean, where it is found today in the sediments. Glacial runoff and melting icebergs lowered the salinity of North Atlantic surface waters, leading to important changes in ocean circulation.

Since both interstadial events and cold periods (including Heinrich events) are registered in the $\delta^{18}O$ of the ice, it is obvious that the climate must have switched between these two modes. Thus, interstadial events were the warm temperature excursions, and the Heinrich events were the periods when Greenland temperatures were at their lowest baseline values. Not all interstadial events were separated by Heinrich events. Only the longest and coldest periods between interstadial events were Heinrich events. With the exception of H2, Heinrich events were immediately followed by interstadial events that were especially warm and long.

Finally, the Antarctic Isotope Maxima, labeled AIM-1 to AIM-12 in figure 10.1, are expressed as positive (warm) $\delta^{18}O$ excursions of the Byrd, EDML, and EDC ice cores. Antarctic Isotope Maxima were, as their name implies, warm periods in Antarctica that are clearly linked to both interstadial events and Heinrich events. Antarctic Isotope Maxima began during the cold periods, including Heinrich events, in Greenland. Antarctic temperatures rose to maxima around the time Greenland temperatures (or $\delta^{18}O$) rapidly increased at the beginning of interstadial events. Then, going forward in time, the $\delta^{18}O$ in Greenland and Antarctica decreased back to baseline values. Atmospheric CO_2 concentrations rose by about 20 ppm during the long AIM's (fig. 10.1). The simplest explanation, given the discussion of atmospheric CO_2 changes in chapter 9, is that increased mixing and ventilation in the Southern Ocean promoted the transfer of CO_2 from the subsurface ocean to the atmosphere. While CO_2 followed Antarctic temperature, as pointed out earlier, CH_4 tracked Greenland temperature.

In this chapter, we will start by discussing, in more detail, the ice core record of climate change in Greenland, and its significance. We will then look at links between Greenland climate and that of extratropical regions. Next, we will consider climate connections between Greenland and Antarctica. In each case, we'll discuss dynamics of climate leading to the observed variations.

CLIMATE CHANGES RECORDED IN THE GREENLAND ICE CORES

The Greenland ice core record of climate is remarkable both for the rapidity of the changes and their magnitude. During the last ice age, rapid climate change events began with warmings that were largely complete within a few decades. Temperature changes were big—glacial-interglacial differences were over 20°C, and rapid warmings encompassed about half this range. What caused these rapid temperature changes in Greenland? As the following discussion shows, the cause seems to have been the extent of wintertime sea ice in the surrounding seas. Climate change around Greenland and the ocean to its south then influenced conditions elsewhere in the Northern Hemisphere and perhaps beyond.

The best-studied rapid warming occurred at 11.6 Ka. This warming ended a cold period in the northern Atlantic region, known as the Younger Dryas, which had interrupted the overall warming associated with the last glacial termination. The warming reached its full magnitude in a couple of decades or less. How can we know this? Two sources of evidence, among others, stand out. First, the warming was associated with an increase in snowfall, marked in the ice cores by a thickening of annual layers. This thickening occurred over only a couple of years (Alley et al. 2003). Second, warming left a remarkable imprint in the isotopic composition of trapped gases. Since heat diffuses slowly, it takes about one hundred years for the warming to infiltrate through the firn—the upper layer of incompletely compacted ice. It is at the base of the firn, lying about 70 m below the surface in Greenland, that gases are trapped as ice grains are sintered together. For a century after a rapid warming, the base of the firn is therefore colder than the surface of the ice sheet. By a process known as thermal diffusion, the heavy isotopes of gases are enriched at the cold end of the temperature gradient, which is to say at the base of the firn. Here, the heavy isotope enrichment is recorded as gases are trapped in isolated bubbles of air. Eventually, the base of the firn warms and the heavy isotope enrichment disappears. Rapid warmings are thus recorded by sharp increases, going forward in time, of the ratio of heavy to light Ar (argon) and N_2 isotopes in the trapped gases, followed by a slower return to background levels (Severinghaus et al., 1998, Grachev and Severinghaus 2005).

Knowing how much temperature rose at the end of the Younger Dryas, we can now estimate the full glacial-interglacial temperature change in Greenland. During this 10°C warming, the $\delta^{18}O$ of ice increased from about $-40‰$ to -36 ‰. Thus, each 1‰ increase in $\delta^{18}O$ of ice reflects a warming of about 2.5°C. The $\delta^{18}O$ increase at the end of the Younger Dryas was only about half of the 8‰ range in $\delta^{18}O$ values that spanned the warming from the height of the last ice age to the Holocene (fig. 10.1). Therefore we infer that, during the height of the last ice age, Greenland was about 20°C colder than at present. This huge temperature change had been independently documented by measuring the temperature of ice in the hole of the GISP2 ice core (central Greenland), down to the bedrock. The current temperature in the deeper part of the ice sheet still "remembers" some of the cold from the last glacial maximum (Cuffey et al. 1995).

The substantial, rapid warmings at the end of the Younger Dryas, and at the start of earlier interstadial events, raise two questions. First, how did Greenland get so cold in the first place? Second, what mechanism is responsible for the warming? The key clue about the origin of these drastic temperature variations comes from data on the lowering of the snowlines of mountain glaciers in Greenland during the last glacial and the Younger Dryas. Snowlines lie approximately at the altitude where summer temperatures are around the freezing point. Denton et al. (2005) observed that the glacial snowlines were about 500–700 m lower than at present. Thus, during the LGM summertime temperatures in

Greenland were only about 5°C colder than today. The question then becomes how annual average temperatures could have been 20°C colder when summertime temperatures were lower by just 5°C. The obvious answer is that wintertime cooling was very large, perhaps 35°C. What could cause Greenland to be so cold in winter? The likely answer is that wintertime sea ice extended across much of the North Atlantic, surrounding Greenland and thermally isolating it from the ocean's heat. Modeling studies support the link between winter sea ice and the very large decreases in Greenland mean annual temperature (e. g., Li et al. 2010).

MILLENNIAL CLIMATE CHANGE IN THE NORTH ATLANTIC DURING THE ICE AGES

The most dramatic imprint of North Atlantic climate change during the last ice age is the presence of discrete sedimentary layers rich in ice-rafted detritus (IRD), each representing a Heinrich event. As described above, these sedimentary layers correspond to the cold periods in Greenland labeled H-1 to H-5 in figure 10.1. The solid black lines in figure 10.2 show the region of maximum IRD accumulation between 13–25 Ka according to Ruddiman (1977) who first discovered the presence of extensive IRD in North Atlantic sediments. Figure 10.2 also shows all the cores in which Heinrich events have been identified, according to Hemming (2004).

Ice-rafted detritus in northeast Atlantic sediments was concentrated in discrete layers that are equivalent

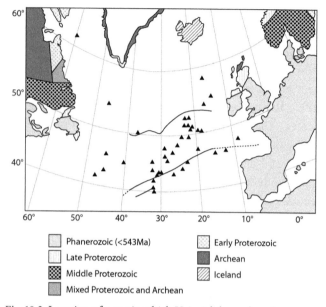

Fig. 10.2. Location of cores in which Heinrich layers have been discovered, according to the review of Hemming (2004). The black lines mark the region of maximum ice-rafted detritus according to Ruddiman (1977). Patterns show ages of rocks for IRD source areas.

in age to the H-events marked in figure 10.1, above the NGRIP $\delta^{18}O$ curve (Heinrich 1988). Most of these layers were composed almost entirely of IRD. The late Gerard Bond, working with colleagues, made detailed studies of the distribution and nature of the IRD. Bond and Lotti (1995) showed that, in addition to layers rich in IRD associated with Heinrich events, there were smaller IRD accumulations associated with the shorter cold events between each of the interstadial events.

Chemical and mineral properties can be characteristic of an IRD source, allowing us to assess its provenance (Hemming 2004). A key indicator of provenance is the abundance of radiogenic isotopes in a rock or mineral. Radiogenic isotopes are stable isotopes produced by the radioactive decay of a "parent" isotope. For example, ^{87}Rb (rubidium) decays to ^{87}Sr, increasing the ratio of ^{87}Sr/^{86}Sr in a rock or mineral (^{86}Sr is one of three stable isotopes of strontium that, unlike ^{87}Sr, has no radioactive source). The older a mineral, and the higher its rubidium concentration relative to strontium, the higher will be its ^{87}Sr/^{86}Sr. In addition to strontium, the isotopic composition of argon, neodymium, and lead varies in minerals for analogous reasons. The isotopic composition of these elements changes in characteristic ways in rocks around the North Atlantic. There are other distinctive features of the composition of ice-rafted detritus: it can contain detrital $CaCO_3$ and $MgCO_3$ mud (from ground-up limestone and dolomite in the source regions), volcanic glasses from Iceland, and quartz grains stained red by Fe_2O_3 that probably come from Greenland. These distinguishing properties allow us to trace minerals making up IRD in North Atlantic sediments back to their source.

Studies of these properties show that the main source of IRD in the Heinrich layers is from the area around the Hudson Straits (fig. 10.2), which drained the region of the Laurentide Ice Sheet centered on Hudson's Bay. This source accounts for much of the coarse-grained material as well as the carbonate mud. When examined in detail, however, there is a rich record of material deriving sequentially from different regions. Heinrich events 3 and 6 are represented by much less accumulation than the other events, and the eastern margin of the Atlantic was an important source. There are differences in the other Heinrich events, as different source regions contribute material at different times. The sediments thus record a complex sequence of iceberg inputs from different areas around the North Atlantic, and show that the sequence of delivery may have changed from one event to the next.

Meltwater from icebergs lowered the salinity of the surface ocean. The clearest evidence comes from data on the $\delta^{18}O$ of planktonic Foraminifera. During Heinrich 4, for example, summertime surface water temperatures (inferred from the species abundances of planktonic forams) cooled by about 3°C, which by itself would have caused the $\delta^{18}O$ of planktonic Foraminifera to increase by about 0.7‰. However, $\delta^{18}O$ actually fell by approximately 1.5‰ in the band of maximum IRD deposition at about 50° N. This decrease, due to the addition of glacial meltwater, corresponds to a salinity drop of about 2–3 units out of 35, which is the average oceanic value. The lowered salinity during Heinrich events caused a drop in the density of surface water in the North Atlantic. This density change had important consequences for ocean circulation and global climate, as we will see below.

CLIMATE CHANGE IN THE TROPICS AND MIDLATITUDES

Climate variations linked to the interstadial events, Heinrich events, and Antarctic isotope maxima have been observed in many extrapolar regions. We focus on four: southeastern China, the currently arid areas of the American West, the tropical Atlantic north of Venezuela, and southeastern Brazil. We start with China, where high resolution speleothem records reveal details of climate change linked to Dansgaard-Oeschger (DO) events.

We discussed records of $\delta^{18}O$ in Chinese caves in the last chapter (see also fig. 9.6). There is variability in $\delta^{18}O$ that is tightly linked to Northern Hemisphere insolation. At times of high insolation, $\delta^{18}O$ of speleothem $CaCO_3$ is low, signaling a change in the hydrologic cycle that most likely involves increased rainfall during the summer monsoon. Inspection of the record shows that there is also millennial variability, reflected in the very noisy records between ~30 and 70 Ka, for example (fig. 9.6). The records for Hulu Cave, southeastern China, are shown on an expanded scale in the top of figure 10.3, with $\delta^{18}O$ decreasing upward (Y. J. Wang et al. 2001). The bottom curve is the GISP2 isotopic temperature record, plotted from 0 to 77 Ka with warmer temperatures toward the top. There is a minimum in speleothem $\delta^{18}O$ for each DO event recorded in the GISP2 ice core. Caves recording precipitation from the Indian monsoon (not shown) follow the same pattern: $\delta^{18}O$ was lower during Dansgaard-Oeschger events in Greenland.

....

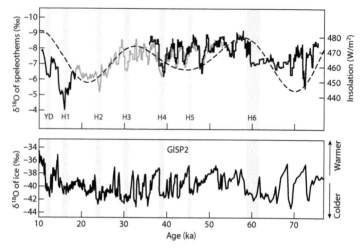

Fig. 10.3. *Top*: The $\delta^{18}O$ of CaCO$_3$ from speleothems in Hulu Cave, eastern China. Lighter values (trending upward) are thought to indicate a stronger summer monsoon. Numbers on the top plot correspond to the numbers of Heinrich events. *Dashed black line*: Northern hemisphere summer insolation. *Bottom*: Isotopic temperature curve for the GISP2 ice core. Vertical gray bars indicate periods of Heinrich events. From Y. J. Wang et al. (2001).

The $\delta^{18}O$ values of these speleothems were originally thought to indicate the strength of monsoon precipitation. The idea was that rains from the summer monsoon had low $\delta^{18}O$ values. Therefore speleothems with low $\delta^{18}O$ values would indicate a strong summer monsoon, while speleothems with high $\delta^{18}O$ values would indicate a weak monsoon (e.g., Y. J. Wang et al. 2001). However, this view is controversial, in part because there are other processes that could cause the ~1‰ variations in $\delta^{18}O$

observed during millennial duration events (Dayem et al. 2010; Pausata et al. 2011). We provisionally follow most recent authors in regarding these changes as evidence for variations in the strength of the East Asian monsoon. In this view, strong Asian monsoons are linked to warm interstadial events in Greenland.

Wallace Broecker has studied climate history in the Great Basin (centered in Nevada but extending into Utah, Idaho, Oregon, and California) for over half a century. The essential method is to use ancient shorelines as a measure of precipitation—higher shorelines mean larger lakes and more rainfall. Dating these shorelines provides a chronology and information about relative precipitation versus time. His recent collaborative work (Broecker et al. 2009) details lake level history during the first half of the last glacial termination with a focus on Heinrich event 1, which dates from about 17.5 to 15.5 Ka. The salient result is that there is a dramatic increase in lake level at 16.4 Ka, coinciding with the second half of Heinrich event 1. At this one time, at least, precipitation was enhanced in the western United States while Greenland was cold, the northern tropics were dry (see below), and wetlands elsewhere were reduced in extent (as judged from the methane concentration).

Millennial scale climate changes are also recorded in the Cariaco Trench, a "hole" in the tropical Atlantic seafloor at 10° N, just north of Venezuela (Peterson et al. 2000). In sediments deposited during the last ice age, bands of light- and dark-colored sediments alternate. The dark bands correspond to times of rapid mud

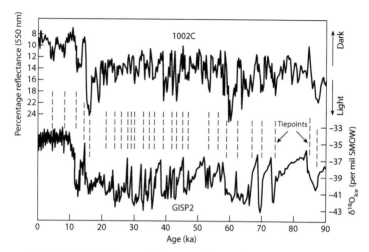

Fig. 10.4. *Top*: Darkness of sediments (lower reflectance) versus time in a Cariaco Trench core north of Venezuela. Darker layers indicate higher concentrations of mud, corresponding to periods of greater precipitation and a higher flux of mud to the study site. *Bottom*: Isotopic temperature curve of GISP2. Tie points connect maxima or minima in the two records. The two curves demonstrate a link between warm temperatures in Greenland and higher precipitation in the tropical Atlantic (Peterson et al. 2000).

accumulation, signifying higher precipitation on the nearby continent. Accumulation of dark bands coincided with Dansgaard-Oeschger events in Greenland (Peterson et al. 2000; fig. 10.4). Thus, precipitation in the northern tropical North Atlantic was elevated during interstadial events, and suppressed during Heinrich events and other cold periods in Greenland.

In several areas of tropical South America, precipitation changes appear to be out of phase with respect to those in the northern tropics, such as the Cariaco Trench. An example comes from northeastern Brazil (10° S), where speleothems grew during wet times and did not grow during dry times. At a particular age, the presence or absence of speleothems provides some indication of relative precipitation. It turns out that, between 10 and 60 Ka, speleothems grew at the times of H1, H4, H5, and H6, and were otherwise dormant. Between 60–90 Ka, the record indicates that speleothems sometimes failed to grow during cold periods in Greenland and dry periods (low reflectance) in the Cariaco Trench (fig. 10.4). This difference in precipitation in the northern and southern tropics is recorded elsewhere and appears to be a robust feature of tropical and subtropical climates during the ice ages.

CLIMATE CHANGE IN ANTARCTICA

We emphasize four salient features of the Antarctic ice core record during the last ice age (fig. 10.1). First, millennial-duration warm events in Antarctica cooccur with interstadial events in Greenland. Second, particularly long and warm Antarctic Isotope Maxima (AIM's) coincide with Heinrich events and the long Greenland interstadials that follow. Third, AIM's begin before the rapid warming in Greenland associated with DO events. Fourth, CO_2 rises by about 20 ppm at the time of H3, H4, H5, and H6. The higher CO_2 concentrations may have been associated with increased winds over the Southern Ocean and more rapid mixing.

CAUSES OF RAPID CLIMATE CHANGE
EVENTS DURING THE ICE AGES

Inspection of figure 10.1 shows that it is very difficult to define a millennial-duration climate cycle as a sequence of events initiated by some identifiable process. For example, an obvious starting point is the Greenland rapid warming, but these are always preceded by warmings in Antarctica. It seems more useful to think about millennial-duration climate change as a repeating chain of events in which the completion of one phase leads to the start of the next. There are many competing ideas about the causes of rapid glacial climate change, but a leading theory is emerging that seems likely to be sustained and expanded. The roots of the hypothesis go back to work of Broecker and collaborators in the mid-1980's and later (e.g., Broecker 1994).

We start with Heinrich events in the North Atlantic. A basic observation is that these events were linked to the presence of large ice sheets in the surrounding land areas, now absent except in Greenland. Source fingerprinting, discussed above, indicates that most of the ice-rafted detritus found in the Heinrich layers came from the Hudson Straits. There are four hypotheses about their origin (Hemming 2004): as either (1) surges of glaciers in Hudson's Bay, (2) surges of ice streams flowing through the Hudson Straits, (3) rapid discharge of large subglacial lakes formed by melting at the bed, or (4) the breaking off of floating ice shelves originating from glaciers that had discharged through the Hudson Straits. In each case, we can think of the system by analogy to a capacitor: ice or basal meltwater reservoirs fill, and then discharge. All four proposals have in common the idea that most ice carrying IRD drained into the North Atlantic through the Hudson Straits, with the accompanying implication that large amounts of freshwater would have been discharged into the North Atlantic in the form of the resulting meltwater.

Accumulation of large amounts of meltwater in the North Atlantic surface, recorded by isotopically light oxygen in planktonic forams (among other measures), will lower the density of surface waters. Many numerical modeling studies have examined the effects of meltwater on ocean circulation and climate (e.g., see the pioneering work of Manabe and Stouffer [1997] and the review by Clement and Peterson [2008]). If the freshwater input is sufficiently great, it will oppose the density increase due to cooling and diminish or stop the sinking of surface waters to depth. Hemming (2004) summarized estimates of the rate of freshwater addition. Some estimates are high enough to cause significant climate and circulation changes, according to modeling studies.

North Atlantic Deep Water (NADW) forms when dense surface waters sink to depth and flow southward, as happens today. In this state, waters lost to sinking must be replaced by tropical waters flowing north. According to modeling studies, in the Heinrich state, introduction of sufficient meltwater cuts off this flow and the warmth it brings to high latitudes. The northern Atlantic region thus cools, to the point where wintertime sea ice can extend south of Greenland, and isolate that landmass.

....

The cooling and growth of sea ice in the North Atlantic leads to climate changes in the regions beyond. When the tropical Atlantic cools, the Intertropical Convergence Zone (ITCZ) shifts southward, along with the belt of tropical precipitation (e.g., Chiang and Bitz 2005; Chiang et al. 2003; Zhang and Delworth 2005). This shift explains both drier conditions recorded on the northern rim of South America (in the Cariaco Trench, about 10° N), and wetter conditions in caves located at about 10° S in northeastern Brazil (X. Wang et al. 2004). In warmer periods, the ITCZ and the locus of precipitation is over the northern coast of South America, leading to wet climates there. During Heinrich events, the ITCZ shifts south, and the northern coast of South America is drier. Further south, there is more precipitation resulting in wetter conditions in northeastern Brazil, for example.

Heinrich events in the North Atlantic induce changes in the Southern Ocean and Antarctica with important consequences. Perhaps most significant is an acceleration in the rate of upwelling and formation of deep water in the Southern Ocean. Two mechanisms lead to this effect. First, the presence of ice in the North Atlantic during Heinrich events drives climate belts to the south, including westerlies blowing over the Southern Ocean. Shifted to the south, they blow over the center of the Southern Ocean and induce upwelling and deep water formation. A symptom of this change is the increase in CO_2 during Heinrich events (e.g., Anderson et al. 2009; Lee et al. 2011). Another aspect is the shift of deep water formation to the south due to the shutdown of NADW. The idea is that deep waters will always form somewhere; warm, shallow waters are mixed down via turbulence into the ocean interior, and the resulting less dense waters will eventually be replaced by denser waters forming in one of the high-latitude regions. If NADW formation is cut off, the locus of deep water production shifts to the Southern Ocean. Ventilation associated with this deep water formation will release CO_2 to the atmosphere (Sigman et al. 2007).

Antarctica warms during Heinrich events (fig. 10.1). The nature of Atlantic Ocean circulation provides an explanation for this surprising discovery. During warm times in Greenland, deep water forms from sinking surface water in the North Atlantic to form NADW. This deep water flows into the South Atlantic and on to the Southern Ocean and the Pacific. The ultimate source of NADW is South Atlantic surface water, which crosses the equator and flows toward the North Atlantic where it is cooled and densified. During the formation of NADW, the North Atlantic thus steals heat from the South Atlantic as the warm water flows north. When NADW formation stops, warm waters remain in the South Atlantic, causing that hemisphere to warm. Related circulation changes in the Southern Ocean concentrate the warming on the continent of Antarctica (e.g., Kageyama et al. 2010).

Heinrich events end because, eventually, fresh meltwater is flushed out of the North Atlantic. NADW production resumes, the North Atlantic warms, and the changes in the tropics and the Southern Hemisphere associated with Heinrich events are reversed. Subsequently, there is a new episode of iceberg or meltwater discharge into the North Atlantic, and the cycle repeats.

What happens during the more frequent climate cycles, involving interstadial events and the intervening cold "stadial" events that occur between the Heinrich events? Our best understanding is that these are muted Heinrich cycles. The interstadial events in Greenland are all associated with interstadial events in Antarctica, and vice versa (fig. 10.1; Wolff et al. 2010). Antarctica warms during cold periods in Greenland, whether or not these cold periods represent Heinrich events. During each event, Greenland warms rapidly when the Antarctic warming reaches its maximum, and then Antarctic and Greenland temperatures fall in concert. Greenland stadials that are not Heinrich events still show elevated levels of ice-rafted detritus, albeit in lower abundances than during Heinrich events (Bond and Lotti 1995).

When it was first understood that rapid $\delta^{18}O$ changes in Greenland ice cores recorded dramatic and fast climate changes in many regions of the globe, there was concern that the modern climate might also change rapidly once a tipping point was crossed. However in studies of the climates of the Holocene and the preceding interglacial (at about 125 Ka), the only evidence for an episode similar to rapid ice age climate change comes from an event dated to 8.2 Ka in which meltwater from the decaying Laurentide ice sheet led to a brief shutdown of NADW formation with attendant consequences to climate. In general, it seems that massive ice sheets are required to produce the rapid climate changes that frequently occurred during the last ice age and beyond.

WHAT CONTROLS THE TEMPO OF DANSGAARD-OESCHGER EVENTS?

One of the most enigmatic observations about interstadial events is their pacing. These events occur with remarkable fidelity on a 1500 year beat, although occasional beats are missed (Alley et al. 2001; Mayewski et al. 1997; Rahmstorf 2002). The question is, what is the cause of this timing? An obvious possibility is that there is a 1500 year cycle in solar intensity. Such a cycle would be imprinted in variations of the ^{10}Be (beryllium) abundance of ice cores and deep-sea sediments. The connection comes from the fact that when the sun shines more brightly, it strengthens the solar magnetic field around the Earth. The stronger field deflects incoming cosmic rays (high energy protons and other nuclei) which undergo nuclear reactions with nitrogen in the atmosphere to produce "cosmogenic" ^{10}Be. This type of ^{10}Be is a radioactive isotope with a half-life of 1.39 Myr and no other significant source. It is removed from air to accumulate in soils, sediments, and polar ice. The absence of a 1500 year cycle in ^{10}Be accumulation (Aldahan and Possnert 1998; Usoskin et al. 2009) suggests that the 1500 year cycles are not driven by changes in solar activity (Bard and Frank 2006). Another alternative is periodicity in glacial discharge stopping the NADW formation, but this process is unlikely to be so highly periodic.

Alley et al. (2001) suggested that the strong periodicity in interstadial events is due to stochastic resonance. In this phenomenon, a system switches between two stable states (or modes) due to the combined effects of a push ("weak directed forcing") and noise, neither of which is strong enough to drive the switch on its own. Friedrich et al. (2010) suggested that the forcing was associated with cooling that would weaken the stratification of the North Atlantic, making circulation in that basin more sensitive to the freshwater balance. Other authors have suggested mechanisms that involve changes in ocean circulation (summarized by Friedrich et al. 2010). We do not yet have a robust understanding of the 1500 year cycles.

In summary, rapid climate change events were persistent features of ice age Earth. These events encompassed a pattern of climate changes stretching from the Arctic to Antarctica, and encompassing temperate latitudes, including monsoonal regions and the tropics as well. Changes in the high latitudes were most strongly manifested in terms of temperature, whereas changes in the middle and low latitudes were manifested largely by variations in precipitation. In general, climate change in the Northern Hemisphere was somewhat out of phase with climate change in the south. Changes in Southern Ocean circulation caused CO_2 to rise when Antarctica warmed, and changes in precipitation intensity caused CH_4 to rise when Greenland was warm. The ultimate causes of rapid climate change events remain uncertain.

REFERENCES

Papers with asterisks are suggested for further reading.

Aldahan, A., and G. Possnert (1998), A high-resolution 10Be profile from deep sea sediment covering the last 70 ka: Indication for globally synchronized environmental events, *Quaternary Geochronology*, *17*, 1023–1032.

*Alley, R. B., S. Anandakrishnan, and P. Jung (2001), Stochastic resonance in the North Atlantic, *Paleoceanography*, *16*(2), 190–198.

*Alley, R. B., J. Marotzke, W. D. Nordhaus, J. T. Overpeck, D. M. Peteet, R. A. Pieldo, R. T. Pierrehumbert, et al. (2003), Abrupt climate change, *Science*, *299*, 2005–2010.

Anderson, R. F., S. Ali, L. I. Bradtmiller, S.H.H. Nielsen, M. Q. Fleisher, B. E. Anderson, and L. H. Burckle (2009), Wind-driven upwelling in the Southern Ocean and the deglacial rise in atmospheric CO_2, *Science*, *323*, 1443–1448.

Bard, E., and M. Frank (2006), Climate change and solar variability: What's new under the sun?, *Earth and Planetary Science Letters*, *248*, 1–14.

Bond, G. C., and R. Lotti (1995), Iceberg discharges into the North Atlantic on millennial time scales during the last glaciation, *Science*, *267*, 1005–1010.

Broecker, W. S. (1994), Massive iceberg discharges as triggers for global climate change, *Nature*, *372*, 421–424.

Broecker, W. S., D. McGee, K. D. Adams, H. Cheng, R. L. Edwards, C. G. Oviatt, and J. Quade (2009), A Great Basin-wide dry episode during the first half of the Mystery Interval? *Quaternary Science Reviews*, *28*, 2557–2563.

Chiang, J.C.H., and C. M. Bitz (2005), Influence of high latitude ice cover on the marine Intertropical Convergence Zone, *Climate Dynamics*, *25*, 477–496.

Chiang, J.C.H., M. Biasutti, and D. S. Battisti (2003), Sensitivity of the Atlantic Intertropical Convergence Zone to Last Glacial Maximum boundary conditions, *Paleoceanography*, *18*. doi: 10.1029/2003PA000916.

Clement, A. C., and L. C. Peterson (2008), Mechanisms of abrupt climate change of the last glacial period, *Reviews of Geophysics*, *46*(RG-4002), 1–39.

Cuffey, K. M., G. D. Clow, R. B. Alley, M. Stuiver, E. D. Waddington, and R. W. Saltus (1995), Large Arctic temperature change at the Wisconsin-Holocene glacial transition, *Science*, *270*, 455–458.

Dayem, K. E., P. Molnar, D. S. Battisti, and G. H. Roe (2010), Lessons learned from oxygen isotopes in modern precipitation applied to interpretation of speleothem records of paleoclimate from eastern Asia, *Earth and Planetary Science Letters*, *295*, 219–230.

Denton, G. H., R. B. Alley, G. C. Comer, and W. S. Broecker (2005), The role of seasonality in abrupt climate change, *Quaternary Science Reviews*, *24*, 1159–1182.

Friedrich, T., A. Timmermann, L. Menviel, O. E. Timm, A. Mouchet, and D. M. Roche (2010), The mechanism behind internally generated centennial-to-millennial scale climate variability in an earth system model of intermediate complexity, *Geoscientific Model Development*, *3*, 377–389.

Grachev, A., and J. P. Severinghaus (2005), A revised +10 +4°C magnitude of the abrupt change in Greenland temperature at the Younger Dryas termination using published

GISP2 gas isotope data and air thermal diffusion constants, *Quaternary Science Reviews*, *24*, 513–519.

Heinrich, H. (1988), Origin and consequences of cyclic ice rafting in the Northeast Atlantic Ocean during the past 130,000 years, *Quaternary Research*, *29*, 142–152.

Hemming, S. R. (2004), Heinrich events: Massive Late Pleistocene detritus layers of the North Atlantic and their global climate imprint, *Review of Geophysics*, *42*(RG1005), 1–43.

Kageyama, M., A. Paul, D. M. Roche, and C. J. Van Meerbeeck (2010), Modelling glacial climatic millennial-scale variability related to changes in the Atlantic meridional overturning circulation: A review, *Quaternary Science Reviews*, *29*, 2931–2956.

Lee, S.-Y., J. C. H. Chiang, K. Matsumoto, and K. S. Tokos (2011), Southern Ocean wind response to North Atlantic cooling and the rise in atmospheric CO_2: Modeling perspective and paleoceanographic implications, *Paleoceanography*, *26*(PA1214), 1–16.

Li, C., D. S. Battisti, and C. Bitz (2010), Can North Atlantic sea ice anomalies account for Dansgaard-Oeschger climate signals? *American Meteorological Society*, *23*, 5457–5475.

Manabe, S., and R. J. Stouffer (1997), Coupled ocean-atmosphere model response to freshwater input: Comparison to Younger Dryas event, *Paleoceanography*, *12*(2), 321–336.

Mayewski, P. A., L. D. Meeker, M. S. Twickler, S. Whitlow, Q. Yang, W. B. Lyons, and M. Prentice (1997), Major features and forcing of high-latitude northern hemisphere atmosphere circulation using a 110,000-year-long glaciochemical series, *Journal of Geophysical Research*, *102*(C12), 26, 345–326, 366.

Pausata, F.S.R., D. S. Battisti, K. H. Nisancioglu, and C. M. Bitz (2011), Chinese stalagmite $\delta^{18}O$ controlled by changes in the Indian monsoon during a simulated Heinrich event, *Nature Geoscience*, 4, 474–480.

Peterson, L. C., G. H. Haug, K. A. Hughen, and U. Rohl (2000), Rapid changes in the hydrologic cycle of the Tropical Atlantic during the last glacial, *Science*, 290, 1947–1951.

*Rahmstorf, S. (2002), Ocean circulation and climate during the past 120,000 years, *Nature*, 419, 207–214.

Ruddiman, W. F. (1977), Late Quaternary deposition of ice-rafted sand in the subpolar North Atlantic (lat 40° to 65° N), *Geological Society of America Bulletin*, 88, 1813–1827.

Severinghaus, J. P., T. Sowers, E. J. Brook, R. B. Alley, and Michael L. Bender (1998), Timing of abrupt climate change at the end of the Younger Dryas interval from thermally fractionated gases in polar ice, *Nature 391*, 141–146.

Sigman, D. M., A. M. de Boer, and G. H. Haug (2007), Antarctic stratification, atmospheric water vapor, and Henrich Events: A hypothesis for Late Pleistocene deglaciations, *Geophysical Monograph Series*, 173, 335–349.

Usoskin, I. G., K. Horiuchi, S. Solanki, G. A. Kovaltsov, and E. Bard (2009), On the common solar signal in different cosmogenic isotope data sets, *Journal of Geophysical Research*, 114(A03112), 1–14.

Wang, X., A. S. Auler, R. L. Edwards, H. Cheng, P. S. Cristalli, P. L. Smart, D. A. Richards, and C.-C. Shen (2004), Wet periods in northeastern Brazil over the past 210 kyr linked to distant climate anomalies, *Nature*, 432, 740–743.

*Wang, Y. J., H. Cheng, R. L. Edwards, Z. S. An, J. Y. Wu, C.-C. Shen, and J. A. Dorale (2001), A high-resolution absolute-dated Late Pleistocene monsoon record from Hulu Cave, China, *Science*, *294*, 2345–2348.

*Wolff, E. W., J. Chappellaz, T. Blunier, S. O. Rasmussen, and A. Svensson (2010), Millennial-scale variability during the last glacial: The ice core record, *Quaternary Science Reviews*, *29*, 2828–2838.

Zhang, R., and T. L. Delworth (2005), Simulated tropical response to a substantial weakening of the Atlantic thermohaline circulation, *Journal of Climate*, *18*, 1853–1860.

11   THE HOLOCENE

···

THE HOLOCENE BEGAN, BY DEFINITION, 11,700 YEARS ago, at the end of the Younger Dryas cold excursion. It extends to the present, although the era beginning around 1850, when man first undertook activities that would have a significant impact on global climate, is sometimes called the Anthropocene. Local and regional climate changes during the Holocene are documented in a very extensive literature. Most regional changes are attributed to variations in Earth's internal climate system (i. e., changes in regional temperature, precipitation, and other climate properties that are not driven by changes in Earth's orbit or solar activity). In this chapter, we focus instead on two modes of climate change that are distinguished by their hemispheric or global imprint. The first group involves climate changes driven by the major insolation changes of the Holocene, the precession-driven shift leading from warmer Northern Hemisphere summers at the start of the epoch to warmer Southern Hemisphere summers today. The second group involves changes linked in some way to variations in the North Atlantic Ocean. This second group has sometimes been attributed to variations in solar luminosity, and we discuss this attribution. Finally, we end with three examples of regional climate change on the continents attributed to internal instabilities, associated in this case with variations in sea surface temperature.

HOLOCENE CLIMATE CHANGE ASSOCIATED WITH PRECESSION

The orientation of Earth's spin axis (box 4) has changed over the past 10 Kyr so that northern summers now occur when Earth is farthest from the sun, whereas at 10 Ka they occurred when Earth was closest to the sun. Northern summertime insolation reached a maximum at about 10 Ka and has declined to the present, when it is near the minimum. This feature is visible on the plot of June insolation versus time before present in figure B, box 4. Wintertime insolation in the Northern Hemisphere, and summertime insolation in the Southern Hemisphere both increase from 10 Ka to today. This change has had two important effects on climate, which is hardly surprising given what we've learned about insolation and climate from Pleistocene studies. First, Northern Hemisphere summers have grown cooler over time, a change recorded mainly by vegetation in the relatively ice-free world of the Holocene. Second, precipitation in the subtropics and tropics has changed progressively during the Holocene. Monsoonal precipitation has grown weaker in the Northern Hemisphere, and the Intertropical Convergence Zone (ITCZ), with its belt of high rainfall, has moved progressively to the south. Of course, climate change may be more complex than suggested by the insolation curves, because two confounding factors come into play. First, a remnant of the Laurentide Ice Sheet survived until about 7 Ka (Carlson et al. 2008), and its high elevation and albedo contributed to cooling meteorologically proximal northern regions (Renssen et al. 2009). Thus, in areas such as western Europe, which are meteorologically tied to eastern Canada, the response to high-latitude insolation was muted at the beginning of the Holocene. In these regions, the Holocene climate optimum occurs between 4–7 Ka rather than earlier. Second, many regions were influenced by more than one mode of climate change, and all relevant modes are conflated in the record.

Summertime temperatures in northern latitudes were higher in the ocean as well as on land. Diatom species abundances and the alkenone temperature index shows that, in the region around Iceland, summer sea surface temperatures were warmer early in the Holocene (Moros et al. 2004; Rasmussen and Thomsen 2010). A record at 60° N, southwest of Iceland, shows the warmest temperatures between about 4–7 ka, followed by a slight cooling (Came et al. 2007).

Most information about high-latitude Holocene land temperatures comes from pollen data, whose proxy quality is based on the fact that plants have optimal temperature ranges. Prentice et al. (2000) summarized a massive amount of pollen data for Europe and North America on three "time slices": today, 6 Ka, and the last glacial maximum. They showed that higher latitudes were warmer in the mid-Holocene. The contrast is strongest in Scandinavia, where mid-Holocene forests are found in regions that host taiga (polar scrub forests) today. The boundary between forest and taiga was also further north in North America during the mid-Holocene. We can get somewhat more information about European climate change by "stacking" pollen records for many different locations (essentially averaging the climate anomalies in all the records), and taking advantage of the fact that pollen gives independent information about summertime temperatures and annual averages. In Scandinavia and northern Germany, July temperatures were warmest around 7–8 Ka and mean annual temperatures were warmest around 6–7 Ka. Mean annual temperatures fell by about 2.5°C after 6 Ka (Seppa et al. 2009). In Europe south of 50°N latitude, there was considerable warming from the beginning of the Holocene to about 8 Ka, after which temperatures were stable. The exception was southeastern Europe (Spain, Portugal, and Italy), which has warmed over the past 8 Kyr (Davis et al. 2003).

In the tropics, Holocene climate change driven by insolation was manifested as variations in the strength of the monsoons, and the migration of the Intertropical Convergence Zone (ITCZ). In some cases, the ITCZ may have aligned with regions affected by monsoon dynamics, with both factors playing a role.

The Intertropical Convergence Zone tends to follow the sun. Intuitively this makes sense, as air will rise where the surface is heated most strongly. During the early Holocene, Earth was closest to the sun in the northern summer; during the later Holocene, Earth was closest to the sun in the southern summer. Accordingly, the summertime position of the ITCZ, and the associated band of rising air and high precipitation, has moved to the south during the Holocene. Areas in the northern tropics that get their precipitation from the ITCZ have become drier, and areas in the southern tropics have become wetter. For example, just north of Venezuela, changes in the composition of sediments in the Cariaco Trench ($11°$N) indicate that precipitation was greatest between 6–10 Ka and diminished progressively toward the present (Haug et al. 2001). In the western tropical Pacific, deep-sea sediment cores centered on the equator indicate that the water has become substantially less salty over the Holocene (Stott et al. 2004). The authors suggested that the changing position of the ITCZ directed water first into the Atlantic, making the Pacific salty earlier in the Holocene, and then later to the Pacific, where the water was freshened.

Insolation also acts on tropical precipitation by influencing the strength of the monsoon. The summer monsoon develops as land is heated more in summertime than the adjacent ocean. Wet air from the ocean flows onto land where it rises, cools, and releases precipitation. When the summer sun shines more brightly, uplift and precipitation are more intense, as discussed earlier. Thus, one expects the monsoon to have been stronger in the Northern Hemisphere at the beginning of the Holocene, and to be stronger in the Southern Hemisphere today. The argument is identical to that in chapter 9 ("The Pleistocene") regarding the relationship between precession and the intensity of Asian summer monsoons as recorded in speleothems.

Not surprisingly, Holocene precipitation responded strongly to precession changes in various monsoon regions. First, consider regions north of the equator. Around the Indian Ocean, there is evidence from both marine and terrestrial deposits for a stronger monsoon earlier in the Holocene. In marine deposits, the evidence comes from the abundance of forams acclimated to upwelling systems (which are more prevalent when monsoons and winds are stronger), from proxy reconstructions of salinity (lower salinity means more rain and river runoff), and from vegetation on nearby lands as recorded by pollen (Anderson and Prell 1993; Gupta et al. 2003; Gupta et al. 2011; Overpeck et al. 1996). On land, evidence for a wet period from about 5 to 10 Ka comes from pollen, lake levels, and other sources (as summarized by Overpeck et al. [1996]). Further support comes from a study of a speleothem in Oman, where $\delta^{18}O$ records the amount of summer monsoon precipitation (Fleitmann et al. 2003). As for other speleothem data, the record is of particular interest because the timescale is very accurate and the record is continuous.

In China, there is also evidence for an intensified East Asian monsoon during the early Holocene. This evidence includes pollen data from southeastern China, data from deposits of loess (windblown dust) showing that there was more intense weathering (hence wetter soils), and proxy salinity data from the South China Sea indicating more runoff (Maher and Thompson 1995; Porter 2001; Zhou et al. 2004). In speleothems from East Asia, $\delta^{18}O$ increases throughout the Holocene toward the present, indicating that the summer monsoon has become progressively weaker (L. Wang et al. 1999; Y. Wang et al. 2005).

South of the equator, in Flores, Indonesia (8°S latitude), summer monsoon intensity increased toward the present (Griffiths et al. 2010). An intriguing example of the influence of precession comes from Lake Challa, at 3°S in Tanzania, where properties of organic chemicals in soils provide an archive of precipitation changes. Lake Challa is close enough to the equator that it will have strong monsoons at times when Earth is close to the sun in northern as well as southern summers. One or the other hemispheres spends summer closest to the sun every 10 Kyr. Consequently, there is an approximate 10 Kyr cycle of precipitation intensity (Verschuren et al. 2009) at this location, as opposed to other monsoon sites where the period of variability is the same as the period of precession (about 20 Kyr).

A particularly interesting example of the influence of insolation on monsoon comes from North Africa. This region is the site of a summer monsoon, with the Atlantic Ocean supplying moisture. The monsoon was much stronger between about 14–5 Ka when summer insolation was of course greater. The consequent increase in precipitation during this interval left its mark on lakes that were higher, more widespread, and fresher than today's; on grasslands and trees in areas that are barren today; on decreases of windblown dust; and on greater river runoff than at present (see summaries, for example, in Liu et al. 2007; Patricola and Cook 2007). Gobero, in central Niger, is extremely arid today. However, from about 9 to 4.5 ka, it hosted a lake that was often up to 5 m deep, and the surrounding lands were arboreal savannah. Local fauna included hartebeest, hippos, and crocodiles. Gobero was occupied for most of the Holocene, as recorded most dramatically by many human skulls and skeletons, but also by tools and implements, jewelry, and (needless to say) garbage (Sereno et al. 2008). Occupation was sparse by 4.5 Ka, and ended by 2.4 Ka.

Another example of wetter early Holocene conditions comes from studies on the distribution of settlements in eastern North Africa (fig. 11.1; Kuper and Kropelin 2006). Prior to 10.5 Ka, known settlements were limited to the Nile River valley. Between that time and 7.4 Ka, people were occupying all of northeast Africa west of the Nile. Between 7.4 and 5.6 Ka, settlements became much more restricted in the region, and of course far more so in the time since 5.6 Ka. These archeological obsevations are just two examples of ways in which orbitally driven climate changes during the Holocene have had a profound impact on human civilization.

MILLENNIAL-DURATION RAPID CLIMATE CHANGE DURING THE HOLOCENE

Studies of ice-rafted detritus in Holocene sediments from the main IRD belt in the North Atlantic suggest that climate changes lasting about a millennium continue to the present, and that these changes may be driven by variability in the brightness of the sun (Bond et al. 2001). There is evidence that these climate changes

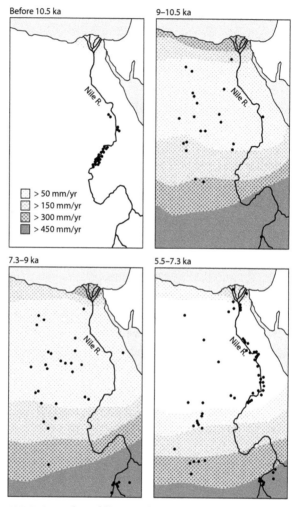

Fig. 11.1. Estimated rainfall in North Africa, and the distribution of human habitation during four time periods of the Holocene (Kuper and Kropelin 2006). Solid dots indicate sites of human settlement, and darker grays reflect more vegetation and higher precipitation.

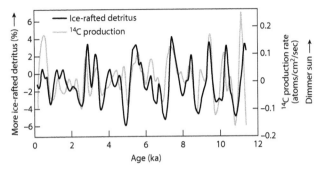

Fig. 11.2. Comparison of Holocene accumulation rates of ice-rafted detritus in North Atlantic sediments, and solar brightness as inferred from past changes in the production rate of ^{14}C by cosmic rays. *Black curve*: Relative accumulation rate of ice-rafted detritus (higher rates toward the top). *Gray curve*: Production rates of ^{14}C versus time (higher rates, toward the top, imply a dimmer sun). From Bond et al. (2001).

in the North Atlantic were synchronous with changes in other regions. We first discuss data for linked millennial-duration climate changes in different regions, and then examine the evidence for solar forcing.

Perhaps the critical data are measurements of the accumulation rates in North Atlantic deep-sea sediments of ice-rafted detritus, with sources in Hudson's Bay, Greenland, and Iceland (fig. 11.2). There are IRD abundance maxima at millennial intervals throughout the Holocene. The maxima are coherent between different IRD minerals in the same core, and they are coherent between cores spanning a wide area (Bond et al. 2001). What is remarkable about this record, however, is that the IRD variations coincide with variations of solar intensity as inferred from the abundance of cosmogenic isotopes. The proxy quality of cosmogenic isotopes was discussed in the last chapter; the bottom line is that, as the sun shines brighter, less radioactive [10]Be and ^{14}C are produced in the atmosphere and contemporaneous sediments. As one example of Bond's results, we compare IRD abundance and cosmogenic isotope production rates in figure 11.2. The left axis is a measure of the average relative abundance of IRD in four North Atlantic sediment cores; the right axis is a measure of [10]Be production rates with respect to the long-term mean. This comparison provides visual evidence, supported by statistics, that ice rafting in the North Atlantic was more rapid when the sun was shining less brightly (Bond et al. 2001).

For two reasons, the Bond curves have become a template against which to compare other high-resolution Holocene records. First, they suggest that solar variability leads to detectable variability in climate that may be manifested elsewhere on the planet. Second, as we saw in chapter 10, ice age climate changes originating in the North Atlantic may be linked to climate change in the tropics and elsewhere. The same links between high- and low-latitude climate would exist in the Holocene, although changes would be smaller and harder to detect. In their extensive review of Holocene climate records from around the globe, Mayewski et al. (2004) identified six periods of anomalous climate in different regions, which they referred to as periods of "rapid climate change." Dates of these periods (9–8 Ka, 6–5 Ka, 3.5–2.5 Ka, 4.2–3.8 Ka, 1.2–1 Ka, and 0.15–0.6 Ka) roughly coincide with cold times in Greenland. On the other hand, correlations between records of Holocene climate changes in other areas are, generally, modest. This is to be expected because Holocene temperature changes were small, and solar forcing was clearly weak and had to compete with other modes of climate variability. These other modes contaminate the record by adding features not originating with the sun, and by sometimes even canceling solar signals. Even in the North Atlantic, Holocene temperature records may look very different from the Bond IRD histories (for example, see Came et al. [2007]).

Solar forcing may have contributed to the two major climate excursions of the last millennium, the Medieval Warm Period and the Little Ice Age (Bard and Frank 2006). The Medieval Warm Period was an interval around 1000 BP when the global climate was somewhat warmer than during the previous millenium. The Little Ice Age was an interval from about 1300 to 1850 when global climate was somewhat cooler. Evidence for these temperature changes comes primarily from tree ring records, data on advances and retreats of mountain glaciers, borehole temperature studies (described in chapter 10), studies of ocean temperatures, and studies of ice-rafted detritus in the North Atlantic (Bond et al. 2001; Bradley 2000; Broecker 2001; Cook et al. 2004; Mayewski et al. 2004; Wanner et al. 2008). In the Bond cosmology, the Medieval Warm Period and the Little Ice Age are the latest warm and cold periods of the current solar cycle, and since about 1850 Earth has been recovering from the Little Ice Age.

WAS SOLAR VARIABILITY AN IMPORTANT CAUSE OF HOLOCENE CLIMATE CHANGE?

The great puzzle of solar forcing is how very small changes in solar output can lead to significant changes in climate. The total variation in solar output estimated for the past millennium, according to one study, is slightly less than 1 W m^{-2} out of 342 W m^{-2}, or about 0.3% (fig. 11.3). Using the Stefan-Boltzmann equation (chapter 1), we can calculate that such a change will lead to a globally averaged temperature change of about 0.2°C. Such a small warming can lead to discernable climate change if climate feedbacks focus the warming in a small area of the planet.

To give some idea about why the North Atlantic region responds so strongly, we briefly summarize results of two atmospheric modeling studies. In the model of Shindell et al. (2001), changes in solar irradiance lead to a series of interactions between the stratosphere and troposphere, and between the high northern latitudes and the tropics. Changes in the UV flux from the sun are about 10 times larger than changes in total solar luminosity. The changing UV flux leads to variations in stratospheric ozone abundance, and UV absorption in the stratosphere then changes. When solar luminosity increases, the tropospheric circulation comes to resemble that of the positive mode of the North Atlantic Oscillation, a measure of the meridional pressure gradient in the North Atlantic. The positive mode is characterized in part by stronger westerly winds. In wintertime,

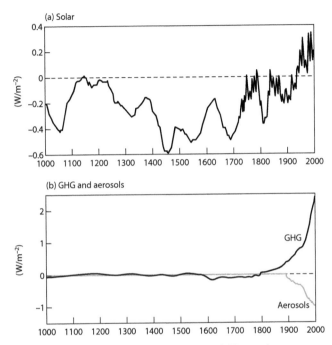

Fig. 11.3. Radiative forcing from the (a) sun and (b) greenhouse gases (*black*) and aerosols (*gray*), estimated for the last millennium (Swingedouw et al. 2011). Variations in solar forcing should be compared with the background value of 342 W m^{-2}.

these winds transport heat from the warm ocean to the cold land, causing warmer continental winters; this is a key process leading to sun-induced climate change. In the model of Swingedouw et al. (2011), the initial consequence of increased solar irradiance comes in the tropics (particularly the tropical Pacific), where solar heating is most intense. Interactions between the atmosphere and the oceans lead to atmospheric pressure changes in high North Pacific latitudes that are then transmitted to the North Atlantic by the jet stream. Changes in sea level pressure are manifested as changes in the North Atlantic Oscillation that then cause temperature changes over land as in the model of Shindell. In addition, changes in ocean temperatures lead to changes in sea ice that act as a positive feedback; warming, for example, melts sea ice, decreasing albedo and increasing heat transfer from ocean to land.

Was solar variability an important cause of Holocene climate variability? We can look at the evidence, but at the present time we just don't know. A number of issues contribute to our uncertainty. First, of course, forcing from the small changes in solar variability is weak, and it is difficult to identify climate records that are dominated by this particular forcing. Second, solar variability is inferred from changes in the abundance of cosmogenic isotopes, but there is controversy about the relation between cosmogenic isotope production rates and solar luminosity. Third, modeling can give guidance about the footprint of climate change that would be caused by the sun, but there is not yet a consensus of modelers as to what this footprint would look like.

OTHER HOLOCENE CLIMATE EVENTS

Many other noteworthy changes in climate occurred during the Holocene, some having profound impacts for civilizations. We focus on three of these events. The first, occuring at 8.2 Ka, was a kind of baby Heinrich event that was a last gasp of the dying glacial world. The other two are representative of a class of events in which important regional precipitation changes on land can be traced back to anomalies in ocean temperature.

The 8.2 Ka event originated with Lake Agassiz, a massive ice-dammed lake in Canada south of Hudson's Bay, which was filled with water from the melting remnants of the Laurentide ice sheet. Ice dams are unstable. When this one gave way, the total volume of the lake— 160,000 km^3 according to one estimate—emptied into the Atlantic within a year or so (Clarke et al. 2003). A low-salinity surface layer formed in the North Atlantic, suppressing formation of North Atlantic Deep Water and causing a regional cooling, as during Heinrich events. The 8.2 Ka event is best recorded by a minimum in the temperatures logged in ice cores from central Greenland (Alley and Agustsdottir 2005). This event is manifested in other areas by climate excursions similar to those characterizing millennial climate change. Glaciers advanced in northern Europe, temperatures fell in ice-free areas to the south, areas affected by the Asian monsoons become drier, and the ITCZ shifted to the south, diminishing precipitation in the low latitudes of the Americas. North American land areas became cooler and drier. A decrease in the atmospheric methane concentration is further evidence for a reduction in Northern Hemisphere precipitation.

Sea surface temperature can influence rainfall in several ways. Sea surface temperatures influence air temperatures, and therefore winds and atmospheric transport. They influence cloudiness and the fluxes of latent heat from low latitudes to high latitudes, and from the sea surface to higher elevations. They also influence the sea-land temperature gradient that drives the summer monsoon. Locally, for example, warmer ocean temperatures will lead to weaker monsoons. The influence of sea surface temperatures on regional precipitation can be assessed based on correlations of data and model properties, but it is not always possible to understand the dynamics underlying a certain relationship. With these complications in mind, two examples illustrate the influence of sea surface temperature on precipitation in the Sahel.

The Sahel is the region, centered around 12° N, stretching across North Africa from the Atlantic to the Indian Ocean, and from the Sahara desert in the north to grasslands and tropical forests in the south. The period from about 1960 to 1980 included very dry intervals, which led to serious famines. In both data and models, drought has been connected to anomalously warm temperatures in the tropical Indian and Pacific Oceans (Biasutti et al. 2008; Giannini et al. 2003). Giannini et al. suggested that warm Indian and Pacific sea surface temperatures would have introduced patterns of atmospheric circulation that intefered with the summer monsoon. Biasutti et al. determined that the meridional sea surface temperature gradient in the Atlantic was also an important influence on precipitation.

"Dust Bowl" refers to a period between 1932 and 1938 when there was a widespread drought in the Great Plains region of the United States, accompanied by intense dust storms. The period was chronicled in John Steinbeck's novel *Grapes of Wrath*, in Walkers Evans' photographs, and in Woody Guthrie's songs ("So long, it's been good to know you/This dusty old dust is a gettin' my home/and I got to be movin' along"). With an approach involving both data analysis and modeling, Schubert et al. (2004) showed that low rainfall was due to a combination of anomalously cool sea surface temperatures in the equatorial Pacific, and warm temperatures in the equatorial Atlantic. Both anomalies influenced upper air circulation in a way that diminished precipitation. The Atlantic effects, which are easier to understand, served to block wet air originating in the Gulf of Mexico from passing onto North America and supplying moisture to the Great Plains. The effects of low precipitation were exacerbated by the drought itself: dry conditions led to low soil moisture and the absence of evaporation that could drive more precipitation.

Many other Holocene droughts have been discovered by paleoclimate studies, including several very severe droughts in the Great Basin over the past millennium. In most cases we don't have a sophisticated understanding of their origin. However, it is likely that many periods of unusually dry (and wet) climates were caused by sea surface temperature anomalies, with dynamics similar to those involved in the Sahel drought and the dust bowl.

REFERENCES

Alley, R. B., and A. M. Agustsdottir (2005), The 8k event: Cause and consequences of a major Holocene abrupt climate change, *Quaternary Science Reviews*, *24*, 1123–1149.

Anderson, D. M., and W. L. Prell (1993), A 300 Kyr record of upwelling off Oman during the late quaternary: Evidence of the Asian Southwest monsoon, *Paleoceanography*, *8*(2), 193–208.

*Bard, E., and M. Frank (2006), Climate change and solar variability: What's new under the sun? *Earth and Planetary Science Letters*, *248*, 1–14.

Biasutti, M., I. M. Held, A. H. Sobel, and A. Giannini (2008), SST forcings and Sahel rainfall variability in simulations of the twentieth and twenty-first centuries, *American Meteorological Society, Journal of Climate*, *21*, 3471–3486.

*Bond, G., B. Kromer, J. Beer, R. Muscheler, M. N. Evans, W. Showers, S. Hoffman, et al. (2001), Persistent solar influence on North Atlantic climate during the Holocene, *Science*, *294*, 2130–2136.

Bradley, R. (2000), 1000 years of climate change, *Science*, *288*(5470), 1353–1355.

Broecker, W. S. (2001), Was the Medieval warm period global? *Science*, *291*(5508), 1497–1499.

Came, R. E., D. W. Oppo, and J. F. McManus (2007), Amplitude and timing of temperature and salinity variability in the subpolar North Atlantic over the past 10 k.y., *Geology*, *35*(4), 315–318.

Carlson, A. E., A. N. Legrande, D. W. Oppo, R. E. Came, G. A. Schmidt, F. S. Anslow, J. M. Licciardi, and E. A. Obbink

(2008), Rapid early Holocene deglaciation of the Lauren-tide ice sheet, *Nature*, *1*, 620–624.

Cook, E. R., J. Esper, and R. D. D'Arrigo (2004), Extra-tropical Northern Hemisphere land temperature variability over the past 1000 years, *Quaternary Science Reviews*, *23*, 2063–2074.

Davis, B.A.S., S. Brewer, A. C. Stevenson, J. Guiot, and D. Con-tributors (2003), The temperature of Europe during the Ho-locene reconstructed from pollen data, *Quaternary Science Reviews*, *22*, 1701–1716.

Fleitmann, D., S. J. Burns, M. Mudelsee, U. Neff, J. Kramers, A. Mangini, and A. Matter (2003), Holocene forcing of the Indian Monsoon recorded in a stalagmite from Southern Oman, *Science*, *300*, 1737–1739.

Giannini, A., R. Saravanan, and P. Chang (2003), Oceanic forcing of Sahel rainfall on interannual to interdecadal time scales, *Science*, *302*, 1027–1030.

Griffiths, M. L., R. N. Drysdale, M. K. Gagan, S. Frisia, J.-S. Zhao, L. K. Ayliffe, W. S. Hantoro, et al. (2010), Evidence for Holocene changes in Australian-Indonesian monsoon rainfall from stalagmite trace element and stable isotope ra-tios, *Earth and Planetary Science Letters*, *292*, 27–38.

Gupta, A. K., D. M. Anderson, and J. T. Overpeck (2003), Abrupt changes in the Asian southwest monsoon during the Holocene and their links to the North Atlantic Ocean, *Nature*, *421*, 354–357.

Gupta, A. K., K. Mohan, S. Sarkar, S. C. Clemens, R. Ravindra, and R. K. Uttam (2011), East-West similarities and differ-ences in the surface and deep northern Arabian Sea records during the past 21 Kyr, *Palaeogeography, Palaeoclimatology, Palaeoecology*, *301*, 75–85.

*Haug, G. H., K. A. Hughen, D. M. Sigman, L. C. Peterson, and U. Rohl (2001), Southward migration of the intertropical convergence zone through the Holocene, *Science*, *293*, 1304–1308.

Kuper, R., and S. Kropelin (2006), Climate-controlled Holocene occupation in the Sahara: Motor of Africa's evolution, *Science*, *313*, 803–807.

Liu, Z., et al. (2007), Simulating the transient evolution and abrupt change of Northern Africa atmosphere-ocean-terrestrial ecosystem in the Holocene, *Quaternary Science Reviews*, *26*, 1818–1837.

Maher, B. A., and R. Thompson (1995), Paleorainfall reconstructions from pedogenic magnetic susceptibility variations in the Chinese loess and paleosols, *Quaternary Research*, *44*, 383–391.

Mayewski, P. A., E.E,R. Rohling, J. C. Stager, W. Karlen, K. A. Maasch, L. D. Meeker, E. A. Mayerson, et al. (2004), Holocene climate variability, *Quaternary Research*, *62*, 243–255.

Moros, M., K. Emeis, B. Risebrobakken, I. Snowball, A. Kuijpers, J. F. McManus, and E. Jansen (2004), Sea surface temperatures and ice rafting in the Holocene North Atlantic: Climate influences on northern Europe and Greenland, *Quaternary Science Reviews*, *23*, 2113–2126.

Overpeck, J. T., D. Anderson, S. Trumbore, and W. Prell (1996), The southwest Indian monsoon over the last 18000 years, *Climate Dynamics*, *12*, 213–225.

Patricola, C. M., and K. H. Cook (2007), Dynamics of the West African monsoon under mid-Holocene precessional forcing: Regional climate model simulations, *Journal of Climate, American Meteorological Society*, *20*, 694–716.

Porter, S. C. (2001), Chinese loess record of monsoon climate during the last glacial-interglacial cycle, *Earth-Science Reviews*, *54*, 115–128.

Prentice, I. C., D. Jolly, and BIOME 6000 Participants (2000), Mid-Holocene and glacial-maximum vegetation geography of the northern continents and Africa, *Journal of Biogeography*, *27*, 507–519.

Rasmussen, T. L., and E. Thomsen (2010), Holocene temperature and salinity variability of the Atlantic water inflow of the Nordic seas, *The Holocene*, *20*(8), 1224–1234.

Renssen, H., H. Seppa, O. Heiri, D. M. Roche, H. Goosse, and T. Fichefet (2009), The spatial and temporal complexity of the Holocene thermal maximum, *Nature Geoscience*, *2*, 411–414.

Schubert, S. D., M. J. Suarez, P. J. Pegion, R. D. Koster, and J. T. Bacmeister (2004), On the cause of the 1930s Dust Bowl, *Science*, *303*, 1855–1859.

Seppa, H., A. E. Bjune, R. J. Telford, H.J.B. Birks, and S. Veski (2009), Last nine-thousand years of temperature variability in northern Europe, *Climate of the Past*, *5*, 523–535.

Sereno, P. C., E.E.A. Garcea, H. Jousse, C. M. Stojanowski, J.-F. Saliege, A. Maga, O. Ailde, et al. (2008), Lakeside cemeteries in the Sahara: 5000 years of Holocene population and environmental change, *PLoS ONE*, *3*(8), 1–22. doi: 10.1371/journal.pone.0002995.

Shindell, D. T., G. A. Schmidt, M. E. Mann, D. Rind, and A. Waple (2001), Solar forcing of regional climate change during the Maunder Minimum, *Science*, *294*, 2149–2152.

Stott, L., K. Cannariato, R. Thunell, G. H. Haug, A. Koutavas, and S. Lund (2004), Decline of surface temperature and salinity in the western tropical Pacific Ocean in the Holocene epoch, *Nature*, *431*, 56–59.

Swingedouw, D., L. Terray, C. Cassou, A. Voldoire, D. Salas-Melia, and J. Servonnat (2011), Natural forcing of climate during the last millennium: Fingerprint of solar variability, *Climate Dynamics*, *36*, 1349–1364.

Verschuren, D., J. S. Sinninghe Damste, J. Moernaut, I. Kristen, M. Blaauw, M. Fagot, and G. H. Haug (2009), Half-precessional dynamics of monsoon rainfall near the East African Equator, *Nature*, *462*, 637–641.

Wang, L., M. Sarnthein, H. Erlenkeuser, P. M. Grootes, J. O. Grimalt, C. Pelejero, and G. Linck (1999), Holocene variations in Asian monsoon moisture: A bidecadal sediment record from the South China Sea, *Geophysical Research Letters*, *26*(18), 2889–2892.

*Wang, Y., H. Cheng, R. L. Edwards, Y. He, X. Kong, Z. An, J. Wu, et al. (2005), The Holocene Asian monsoon: Links to solar changes and North Atlantic climate, *Science*, *308*, 854–857.

Wanner, H., J. Beer, J. Butikofer, T. J. Crowley, U. Cubasch, J. Fluckiger, H. Goosse, et al. (2008), Mid- to late-Holocene climate change: An overiew, *Quaternary Science Reviews*, *27*, 1791–1828.

Zhou, W., X. Yu, A.J.T. Jull, G. Burr, J. Y. Xiao, X. Lu, and F. Xian (2004), High-resolution evidence from southern China of an early Holocene optimum and a mid-Holocene dry event during the past 18,000 years, *Quaternary Research*, *62*, 39–48.

12 ANTHROPOGENIC GLOBAL WARMING IN THE CONTEXT OF PALEOCLIMATE

OUR CURRENT UNDERSTANDING IS THAT, DURING AL-most all of Earth's history, interactions between Earth's interior, surface processes, and global climate feedbacks regulated atmospheric greenhouse gas concentrations so that temperatures were in the habitable range over most of the planet. Certainly this statement is true for the last 600 Myr, while Earth was inhabited by animals; uncertainty exists for most earlier times because we can't accurately characterize surface temperatures.

Evidence of ice in the tropics suggests that the climate went haywire at about 2.4 Ga (2.4 billion years ago), when atmospheric O_2 levels rose, and again about 710 and 630 Ma during Snowball Earth. Otherwise there is little evidence for great ice ages except for the Late Paleozoic (about 360–270 Ma) and the last half of the Cenozoic (34 Ma to the present). Like so many statements about past climate, some qualification is necessary; there were isolated glacial events earlier in the Paleozoic, and nearshore successions of sediment facies reflecting different water depths suggests that, at times in the Precambrian, sea levels changed as ice sheets waxed and waned.

The Late Paleozoic ice ages were followed by a long, ice-free period culminating in the equable climates of the Cretaceous and early Cenozoic. After a period of maximum warmth at around 50 Ma, the planet progressively cooled. Seminal events in this cooling occurred at 34 Ma (development of large ice sheets around Antarctica), 3 Ma (development of large ice sheets on the Northern Hemisphere continents and the origin of 40 Kyr ice age cycles), around 1 Ma (origin of the 100 Kyr ice age cycles), and 10 Ka (the end of the last glacial termination and the rise of atmospheric CO_2 to "preindustrial" levels).

An interesting feature of the Cenozoic climate is that the amplitude of climate change became larger as the mean climate became cooler. This pattern has two main causes. First, when the climate cooled below a certain threshold, large ice sheets developed that led to additional cooling from the ice albedo. Second, there were positive feedbacks between the physical climate system and the atmospheric CO_2 concentration that enhanced the magnitude of glacial-interglacial cycles; cooling climates led to changes in ocean circulation that caused the CO_2 concentration to fall. One could imagine a world with different continental positions where cooling would cause changes in ocean circulation leading to a higher CO_2 concentration. In this case, CO_2 would provide a negative feedback on glacial-interglacial climate change, and climate would be locked into some (presumably cool) intermediate state. Such locking probably occurred at times in the deep past but this phenomenon would be difficult to recognize in the sedimentary record.

Our ancestors gradually evolved in a world of changing climate. Human ancestors split from chimpanzees, our closest relatives, at around 7 Ma. Over the following millions of years, the bodies and brains of our ancestors became larger, bipedality developed, and they gradually evolved into us. The details are controversial. Many anthropologists would say that Neandertals appeared around 250 Ka, and anatomically modern humans developed sometime after 200 Ka. In deposits dating back to about 30 Ka, beautifully sculpted tools and spectacular carvings and cave paintings indicate that *Homo sapiens* had faculties fully equivalent to our own. By about 25 Ka anatomically modern man had interbred with Neandertals, or replaced them in their last European redoubts.

Climate change had a profound influence on human migrations. For example, around 14 Ka, the region of the Bering Straits became habitable enough that Asians could cross over into North America before sea level rose enough to cover the land bridge. The peopling of the Americas was soon followed by the extinction of most large American mammals. The cause(s) of this megafaunal extinction is contentiously debated. One contingent invokes human hunting to extinction. A second view is that the megafauna died because of disease linked to human habitation. Common to these hypotheses is the view that climate change in some way played a large role in the chain of events leading to the mass extinction, as well as the migration to the Americas.

Humans and human ancestors, like all other animals, evolved physically and culturally in a way that was shaped by climate. Our present physiognomy must reflect the long Cenozoic climate deterioration and the evolutionary pressures of the large glacial-interglacial cycles. There is a considerable literature dealing with human origins and climate, but it is difficult to attribute specific biological characteristics to specific climate phenomena.

Around 10 Ka, our ancestors discovered farming, setting the stage for large communities, division of labor, tyranny, and the development of powerful political entities. There seem to have been two requirements for the development of agriculture. The first was the development of the necessary level of intelligence. It seems reasonable to speculate that, by about 30 Ka, anatomically modern humans had evolved the capabilities for the task. The second was the rise of atmospheric CO_2 to the preindustrial level; this occurred by about 10 Ka. This change may have been essential for agriculture to become sustainable. The growth of plants can be limited by the availability of CO_2 in the atmosphere. The crop yield might have allowed farming to compete with hunting and gathering only when CO_2 had risen to preindustrial levels. Subsequently, advances in agriculture allowed larger and larger political entities to develop, leading to the modern world.

Climate has shaped civilization in many ways. Here is one obvious example: fertile areas of the planet are more heavily populated than deserts and ice-covered regions. However, there are many more subtle examples.

Civilization has developed to the point where humankind is now interacting with climate rather than merely responding. The primary way we are doing this is by adding CO_2, the leading agent for natural climate change over the Phanerozoic and beyond. We can assess the effects of this action using models that predict future climate, and by intelligently judging the implications of the paleoclimate record. Our best understanding is that fossil fuel emissions will lead to global warming, sea level rise, and large regional changes in rainfall during the coming centuries. With a high probability, we are already experiencing these effects in our climate. Already the planet has warmed by nearly 1°C. Sea level is rising at the rate of about 30 cm/century, and this rate is likely to increase as the planet warms.

How should we view the prospect of anthropogenic climate change? From the perspective of paleoclimate, it might not be particularly troubling, or even seem unwelcome. The present world is good enough for human habitation. However, it would improve if Greenland and Antarctica were unglaciated and habitable, and if there was more rainfall in areas that are currently deserts. For humans, in other words, the world might be more habitable if conditions resembled the high CO_2 equable climates of the Cretaceous, Paleocene, and Eocene.

The problem of anthropogenic global change, then, is not necessarily that we are heading for a less habitable planet. The problem is that both natural ecosystems and civilizations are aligned to the historic pattern of climate and water resources. Global warming will destroy this alignment is some regions. The most obvious example is sea level rise, which will render regions uninhabitable that are now occupied by tens or hundreds of millions of people. Shifting temperatures and rainfall belts will open some northern areas to agriculture while making agriculture impossible in some currently farmed regions. The disappearance of mountain glaciers will make water unavailable for agriculture in the seasons it is needed, and will supply water at other times when it may not be used efficiently.

The continued burning of fossil fuels will cause the atmospheric CO_2 concentration to rise. If we burn all readily available fossil fuels in the next few hundred years, we are likely to drive the atmospheric CO_2 concentration up to 1500 ppm or so, over five times the preindustrial level. This estimate takes into account that over hundreds of years, a large fraction of CO_2 is taken up rather quickly by the growth of forests and by dissolution in the oceans.

This high atmospheric CO_2 level would be unsustainable. The warm temperature and high CO_2 burden mean that once fossil fuels were exhausted, weathering would consume CO_2 faster than it is added by natural sources. The excess CO_2 would thus be slowly consumed and dissipated by weathering, exactly as for the Paleocene-Eocene Thermal Maximum. Carbon dioxide concentrations would fall, over a period of about 100,000 years, back toward their natural equilibrium level. With such a long horizon, two other factors would come into play. The first is orbital change and the natural climate cycle, which would push Earth back toward a glacial mode at some point. The second is additional transformations of the environment by humans. These tranformations are likely to be severe but cannot now be predicted. As for the past 4.5 Ga, Earth's climate in the near geologic future will be determined by changes in greenhouse gases, albedo, Milankovitch forcing, and perhaps solar variability. However, we cannot now know the forcings that will dominate climate change hundreds of thousands of years or more in the future, and hence cannot judge how climate will respond. Stay tuned.

Glossary

Albedo—Planetary reflectance; the fraction of sunlight incident on Earth that is reflected directly back to space.

Alkenones—See Uk_{37} index.

Antarctic isotope maxima—Relatively warm periods in Antarctica, recognized from the isotopic composition of ice in ice cores, which occurred during the recent ice ages. Antarctic isotope maxima are associated with increasing atmospheric CO_2 concentrations and very cold periods in Greenland and surrounding areas.

Aragonite seas—The state of the oceans at those times during the last 543 Myr when the Mg^{2+}/Ca^{2+} ratio of seawater was high, causing the inorganic $CaCO_3$ precipitating from seawater to crystallize as the mineral aragonite.

Biological pump—The transfer of carbon from the surface ocean to the deep sea by the formation and sinking of organisms and biological debris.

C3 and C4 grasses—Plants distinguished by the manner in which they acquire CO_2 for transformation into tissue. C3 plants transform atmospheric CO_2 directly into a 3-carbon compound. C4 plants add atmospheric CO_2 to a 3-carbon compound to make a 4-carbon compound and then release the CO_2 molecule in the proximity of the enzyme Rubisco for photosynthesis. Some grasses are C4; all other plants are C3.

Compensation depth—The depth below the sea surface at which all accumulating $CaCO_3$ dissolves.

Coriolis force—An apparent force associated with Earth's spin that influences the motion of fluids.

Banded iron formation—A sedimentary sequence consisting of alternating layers of iron oxide and chert.

Brine inclusion—A sample of evaporating seawater trapped by precipitating salts.

Calcite seas—The state of the oceans at those times during the last 543 Myr when the Mg^{2+}/Ca^{2+} ratio of seawater was low, allowing inorganic $CaCO_3$ precipitating from seawater to crystallize as the mineral calcite.

Climate sensitivity—The rise in average global temperature associated with a doubling of the atmospheric carbon dioxide concentration.

Coccolithophorids—Single-celled photosynthesizing organisms with external skeletons made up of plates, or liths, of $CaCO_3$.

Cyclothems—Repetitive sedimentary sequences formed from rising and falling sea levels resulting from the melting and regrowth of ice sheets.

δ notation—The notation used to report the relative abundances of the stable isotopes of an element. A δ value of 1 per mil signifies a difference in abundance ratios of 1 part per thousand.

Degassing—In a global geochemical context, the transfer of volatile elements and compounds (CO_2, water, noble gases, etc.) from the hot solid Earth to the oceans and atmosphere.

Diagenesis—Chemical alteration of fossils or inorganic sediment after their initial deposition.

Diamictites—Piles of unsorted rocks and dirt. Some diamictites have a glacial origin.

Dissolved inorganic carbon—Carbon dioxide, carbonic acid, bicarbonate, and carbonate. The concentration of dissolved inorganic carbon is the sum of the concentration of these four species.

Dropstones—Rocks released by melting icebergs.

Eccentricity—A measure of the "unevenness" or noncircularity in the path of Earth's orbit around the sun.

El Niño—A condition in the equatorial Pacific Ocean in which surface waters are unusually warm in the eastern side. During El Niño events, upwelling is suppressed and patterns of precipitation and winds are altered in the tropics and beyond.

Equable climates—The term describing periods of climate much warmer than the present, particularly in high latitudes.

Forcing or climate forcing—An influence leading to a change in Earth's climate. Examples include changes in greenhouse gas concentrations or changes in solar luminosity.

40K world—The period extending from about 2.5 Ma to 1 Ma, when Earth's climate was characterized by glacial-interglacial cycles of 40 Kyr duration, driven by changes in obliquity.

Foraminifera—A phylum of amoeboid protists with $CaCO_3$ skeletons that is ubiquitous in the ocean surface and sea floor. Foraminifera are common in ocean sediments and extensively studied for the information they contain about past climates. Planktonic foraminifera live near the sea surface and benthic foraminifera live on the sea floor.

Geocarb model—A mathematical model advanced by R. A. Berner and colleagues to account for major geological and biological processes that influence the atmospheric CO_2 concentration and to reconstruct its value during the last 543 Ma.

Gigaton (Gt)—As used here, a measure of the mass of carbon in a reservoir, corresponding to 10^9 tons or 10^{15} grams.

Glacioeustacy—Changing sea levels resulting from the growth and decay of continental ice sheets.

Heinrich events—Intervals of massive iceberg discharge into the North Atlantic. Melting of these icebergs led to major changes in ocean circulation and regional climates.

Ice cores—Cores of ice obtained by drilling through glaciers, and analyzed for various chemical and isotopic properties that reflect past climates.

Ice-rafted detritus (IRD)—Small rock fragments that are entrained by glaciers, transported to distant areas of the ocean in icebergs, and released upon melting.

Interplanetary dust particles—Fine solar system debris slowly raining down on Earth.

Interstadial events—Warm periods in the North Atlantic, with correlative climate changes elsewhere, and associated with rapid climate changes during glacial times. Interstadial events are also known as "Dansgaard-Oeschger events."

Isostacy—The depression or emergence of bedrock in response to the change in a load, such as the growth or decay of a large ice sheet.

Isotope fractionation—The partial separation of isotopes in chemical reactions or physical processes.

Isotopes—Forms of an element distinguished by the number of neutrons in the nucleus. Stable isotopes maintain their integrity indefinitely. Radioactive isotopes transform at a characteristic rate into other elements.

Lapse rate—The rate at which temperature decreases with elevation above Earth's surface.

Lysocline—The depth below the sea surface at which $CaCO_3$ in surface sediments first starts to dissolve.

Meridional heat transport—The transfer of heat from low latitudes to high latitudes by the movement of air or seawater.

Methane hydrates—Solid compounds, formed under high pressure in sediments, composed of water and methane.

Methanogenesis—The biological process in which organic carbon is decomposed to methane and carbon dioxide to produce energy.

Milankovitch theory—A theory that attributes glacial cycles to changes in seasonality associated with variations in obliquity, eccentricity, and precession of Earth's orbit around the sun.

Mobile reservoirs—Surficial reservoirs in which carbon is exchanged over relatively short time scales. They include the land biosphere, soil, atmosphere (CO_2), and seawater (containing dissolved inorganic carbon).

Monsoon—The period of heavy summer rainfall in certain continental regions induced by heating of the land, and leading to rising air and precipitation.

North Atlantic Deep Water (NADW)—A major oceanic water mass that forms when surface water cools in the North Atlantic, becomes dense, sinks, and flows southward; it accounts for a significant amount of the interior waters in the Atlantic

Northern Hemisphere glaciation—An imprecise term generally used to describe the appearance of large ice sheets in North America and Eurasia.

Obliquity—The inclination of Earth's spin axis with respect to a line perpendicular to the plane of planets' orbits around the sun.

100K world—The last ~1 Mya of Earth history, characterized by large glacial-interglacial cycles of 100 Kyr duration.

Paleosol—A fossil soil.

PDB—The arbitrarily selected $CaCO_3$ standard against which the $\delta^{18}O$ and $\delta^{13}C$ of $CaCO_3$ are measured and reported. If a sample has a δ value of zero, its isotopic ratio is identical to PDB; if its δ value is greater than zero, the heavier isotope is more abundant in the sample than in PDB.

Phytoplankton—Single-celled photosynthesizing organisms responsible for most organic matter production in the open ocean.

Polymorph—Forms of a single chemical compound distinguished by the arrangement of atoms in the crystal.

Precession—The very slow wobble of Earth's spin axis as it orbits the sun.

Preindustrial time—The period before industrial activities began affecting the atmospheric CO_2 concentration, generally taken as preceding 1800 or 1850.

Radiative equilibrium—The condition in which any body in a vacuum (including Earth) is gaining and losing heat at the same rate.

Recrystallization—The process by which crystals dissolve and reprecipitate. The primary chemical signal is generally lost in the process.

Rapid climate change events—A descriptive term that is also used to specifically represent events during glacial periods, lasting hundreds or thousands of years, in which Greenland and surrounding areas warmed very quickly, and climates changed simultaneously over many regions of the planet.

Rhythmites—Cyclic sediments formed by repeating events.

Snowball Earth events—Periods when Earth is believed to have been completely covered by ice.

Snowline—The elevation above which the ground is covered by a glacier during the entire year.

Solar luminosity—The rate of energy generation by the sun.

Speleothems—Deposits of $CaCO_3$ formed from cave waters, including stalagmites, stalactites, and flowstones.

Stratosphere—The region of the atmosphere immediately above the troposphere, extending from about 15 to 50 km in elevation, and characterized by increasing temperature with elevation.

TEX$_{86}$ index—An index that reflects the relative abundance of specific, similar, organic compounds (membrane lipids). These compounds are produced by microorganisms, and their relative abundance is sensitive to the temperatures at which the organisms grow. Measuring the index in organic debris of ancient sediments allows us to reconstruct growth temperatures, and hence water temperatures, at a time in the past.

Thermocline—A depth interval in the oceans in which seawater temperature decreases rapidly with depth, and density increases rapidly.

Tropopause—The boundary between the troposphere and the stratosphere, lying about 15 km above Earth's surface.

Troposphere—The lower 15 km or so of Earth's atmosphere, characterized by a decrease in temperature with elevation.

Uk37 index—An index analogous to TEX$_{86}$ that reflects the relative abundance of specific, similar, organic compounds (alkenones). These compounds are produced by microorganisms, and their relative abundance is sensitive to the temperatures at which the organisms grow. Measuring the index in organic debris of ancient sediments allows us to reconstruct growth temperatures, and hence water temperatures, at a time in the past.

Weathering—The chemical process in which igneous and metamorphic rocks on Earth's surface react with CO_2. The rocks are partly dissolved and clay minerals are left as residues, comprising soils.

Index